Philosophy and Logic of Predication

STUDIES IN PHILOSOPHY OF LANGUAGE AND LINGUISTICS

Edited by Piotr Stalmaszczyk

VOLUME 7

Advisory Board:
Emma Borg (University of Reading)
Manuel García-Carpintero (University of Barcelona)
Hans-Johann Glock (University of Zurich)
Paul Livingston (University of New Mexico)
Joanna Odrowąż-Sypniewska (University of Warsaw)
Maciej Witek (University of Szczecin)
Marián Zouhar (Slovak Academy of Sciences, Bratislava)

Piotr Stalmaszczyk (ed.)

Philosophy and Logic of Predication

Bibliographic Information published by the Deutsche Nationalbibliothek
The Deutsche Nationalbibliothek lists this publication in
the Deutsche Nationalbibliografie; detailed bibliographic
data is available in the internet at http://dnb.d-nb.de.

Library of Congress Cataloging-in-Publication Data
Names: Stalmaszczyk, Piotr, editor.
Title: Philosophy and logic of predication / [edited by] Piotr Stalmaszczyk.
Description: New York : Peter Lang, 2017. | Series: Studies in philosophy of
  language and linguistics, ISSN 2363-7242 ; Vol. 7 | Includes index.
Identifiers: LCCN 2016052514 | ISBN 9783631669204
Subjects: LCSH: Predicate (Logic)
Classification: LCC BC181 .P45 2017 | DDC 160--dc23 LC record available at
https://lccn.loc.gov/2016052514

This publication was financially supported by the Institute of English Studies
and Faculty of Philology of the University of Łódź.

Cover image: © Ingae / Fotolia.com

ISSN 2363-7242
ISBN 978-3-631-66920-4 (Print)
E-ISBN 978-3-653-06448-3 (E-PDF)
E-ISBN 978-3-631-70654-1 (EPUB)
E-ISBN 978-3-631-70655-8 (MOBI)
DOI 10.3726/b10706

© Peter Lang GmbH
Internationaler Verlag der Wissenschaften
Frankfurt am Main 2017
All rights reserved.
Peter Lang Edition is an Imprint of Peter Lang GmbH.

Peter Lang – Frankfurt am Main · Bern · Bruxelles · New York ·
Oxford · Warszawa · Wien

All parts of this publication are protected by copyright. Any
utilisation outside the strict limits of the copyright law, without
the permission of the publisher, is forbidden and liable to
prosecution. This applies in particular to reproductions,
translations, microfilming, and storage and processing in
electronic retrieval systems.

This publication has been peer reviewed.

www.peterlang.com

Table of contents

Contributors 7

Piotr Stalmaszczyk
Predication Theory: Introduction 9

Part I. From Aristotle to the Future

António Pedro Mesquita
Aristotle on Predication 25

Nino B. Cocchiarella
On Predication, A Conceptualist View 53

Ignacio Angelelli
The Impact of Traditional Predication Theory on the Notion of Class 93

Allan Bäck
The Future of Predication Theory 101

Part II. Philosophy and Beyond

Mieszko Tałasiewicz
The Sense of a Predicate 129

Danny Frederick
The Unsatisfactoriness of Unsaturatedness 145

David Liebesman
Sodium-Free Semantics: The Continuing Relevance of
The Concept *Horse* 167

Peter Hanks
Predication and Rule-Following 199

Bjørn Jespersen
Is Predication an Act or an Operation? 223

Jacek Paśniczek
Meinongian Predication. An Algebraic Approach 247

Index 271

Contributors

Ignacio Angelelli
The University of Texas at Austin, USA
angelelli@utexas.edu

Allan Bäck
Kutztown University, Pennsylvania, USA
back@kutztown.edu

Nino B. Cocchiarella
Indiana University, USA
cocchiar@indiana.edu

Danny Frederick
Slate House, Hunstan Lane, Old Leake, Boston, UK
dannyfrederick77@gmail.com

Peter Hanks
University of Minnesota, USA
pwhanks@umn.edu

Bjørn Jespersen
VSB-Technical University of Ostrava, Czech Republic
bjorn.jespersen@gmail.com

David Liebesman
University of Calgary, Canada
david.liebesman@gmail.com

António Pedro Mesquita
University of Lisbon, Portugal
apmesquita@netcabo.pt

Jacek Paśniczek
Maria Curie-Skłodowska University, Lublin, Poland
jpasnicz@bacon.umcs.lublin.pl

Piotr Stalmaszczyk
University of Łódź, Poland
piotrst@uni.lodz.pl

Mieszko Tałasiewicz
University of Warsaw, Poland
m.talasiewicz@uw.edu.pl

Piotr Stalmaszczyk
University of Łódź

Predication Theory: Introduction

It is no news that a theory of predication is no sooner formulated than it generates puzzles.
Wilfrid Sellars "Towards a Theory of Predication" (1985)

1. Predication Theory

Research in linguistics, philosophy of language, ontology, metaphysics, and logic makes ample, both explicit and implicit, use of the concept of predication. In the words of Ignacio Angelelli predication has been and continues to be the central topic of logic and ontology (or metaphysics) in the entire history of philosophy, from Antiquity till now (Angelelli, this volume); whereas for Nino Cocchiarella "predication has been a central, if not the central, issue in philosophy since at least the time of Plato and Aristotle" (Cocchiarella 1989: 253), and for John Searle, within a different theoretical and methodological frame, "predication, like reference, is an ancient (and difficult) topic in philosophy" (Searle 1969: 97). The following overview very briefly mentions some of the most important approaches to predication and predication theory, especially in the context of the articles collected in this volume.

In traditional grammar, predication is the relation between the subject and the predicate. In logic, predication is the attributing of characteristics to a subject to produce a meaningful statement combining verbal and nominal elements. This understanding stems from Aristotelian logic, where the term (though not explicitly used by the philosopher) might be defined as "saying something about something that there is".[1]

In logical inquiries, this classical definition is echoed in more recent investigations by the 'thing-property' relation holding in appropriate propositions (e.g. in Reichenbach 1947). According to Quine (1960), predication is the basic combination in which general and singular terms find their contrasting roles:

---

1   This definition may be inferred from Aristotle's concept of a proposition, understood as a "statement, with meaning as to the presence of something in a subject, or its absence, in the present, past, or future, according to the division of time" (*On Interpretation*, 17a23). For a discussion of predication and predication types in Aristotelian theory see Moravcsik (1967), Lewis (1991), Bäck (2000, esp. Ch. 6), and Mesquita (this volume).

"Predication joins a general term and a singular term to form a sentence that is true or false according as the general term is true or false of the object, if any, to which the singular term refers" (Quine 1960: 96); Quine also considers predication to be one of the mechanisms which join occasion sentences. Quine's treatment of predication is characteristic of most logical approaches in being bipartite, i.e. it focuses on the (contrasting) roles of two elements in the relation (disregarding the importance of a possible triggering operator), it also distinguishes monadic and polyadic predication.

The above ideas are close to Lorenzen's (1968) observation concerning 'basic statements' (*Grundaussagen*), the simplest structures of a language which are composed of a subject and a predicate. Strawson (1971) stresses that predication is an assessment for truth-value of the predicate with respect to the topic, similarly Searle (1969), who also provides a speech act theory of predication, observing that it is not a separate speech act at all.[2] According to Link (1998) predication is the basic tool for making judgments about the world, it relates predicates and individual terms to form sentences, semantically "that means that some individuals are said to have properties or to stand in relation to one another" (Link 1998: 275).[3]

Predication can be also understood as the nexus between a subject and a predicate expression (Chierchia 1985; Cocchiarella 2013), the basis of the unity of a speech act (Cocchiarella 2013; this volume). For Victor Dudman "a theory of predication for a language is meant to explain what can be said in it, and how" (Dudman 1985: 43). On the other hand, Sellars (1985, §83) stresses the necessity of distinguishing "the *psychological* and the *logical* dimensions of predication), and Cocchiarella (2013: 302) distinguishes "the nexus of predication in reality that is part of this natural realism from the nexus of predication in our speech and mental acts".[4]

The traditional grammatical approach to predication, with the division of the clause (or simple sentence) into two basic constituents: *subject* and *predicate*,

---

2   Searle claims that predication is "an abstraction, but it is not a separate act. It is a slice from the total illocutionary act; just as indicating the illocutionary force is not a separate act, but another slice from the illocutionary act", and that predication "presents a certain content, and the mode in which the content is presented is determined by the illocutionary force of the sentence" (Searle 1969: 122).
3   Link further observes that in philosophical terms "that corresponds to the traditional distinction between particulars and universals, brought together through the instantiation relation" (Link 1998: 275).
4   For some recent critical discussion on the relation between predication and purported mental processes, see Peregrin (2011).

can be seen in, among others, Bloomfield's (1933) analysis of 'favorite sentence-forms', and Hockett's (1958) operation of conjoining. However, Jespersen (1937) abandoned the term altogether, and instead introduced the concept of *nexus*, a relation joining two ideas, justifying his decision in the following way:

> It would probably be best in linguistics to avoid the word predication altogether on account of its traditional connexion with logical theories. In grammar we should, not of course forget our logic, but steer clear of everything that may hamper our comprehension of language as it is actually used; this is why I have coined the new term nexus with its exclusive application to grammar (Jespersen 1937: 120).

Terminological considerations notwithstanding, Jespersen's conception of predication is explicitly bipartite and firmly relational: a predication is established by some specific combination of what becomes the subject and the predicate, or, in somewhat looser terms, "something happens in a nexus" (Jespersen 1937: 121).

Krifka (1998) claims that a predication establishes a relation of a specified type between a number of parameters, or semantic arguments. For example, sentences with intransitive verbs establish a relation that holds of the subject for some event, and sentences with transitive verbs establish a relation that holds between the subject, the object, and some event.

In Davidson's approach to verb semantics predication can be specified as a relation between a verb and one of its semantic entailments (Davidson 1967), or as a combinatory relation which makes it possible to join a property and an argument of the appropriate semantic type to form a formula whose truth or falsity is established according to whether the property holds of the entity denoted by the argument or not. Elsewhere Davidson related predication to the problem of 'the unity of proposition' and claimed that "(…) if we do not understand predication, we do not understand how any sentence works, nor can we account for the structure of the simplest thought that is expressible in language" (Davidson 2005: 77).[5]

Early generative grammar, in the Chomskyan tradition, paid very limited attention to the notion, and later studies, e.g. Williams (1980) and Rothstein (1985, 1992), viewed predication as a primitive syntactic relation. Whereas Williams argued for an indexing approach and almost equaled predication with semantic role assignment, Rothstein claimed that the semantic and syntactic concepts of predication are distinct, and the relation which holds between predicates and

---

5   For a critical discussion of this claim, see Peregrin (2011). For some recent studies investigating, from different angles, predication and the unity of the proposition, see Gaskin (2008), García-Carpintero (2010), Collins (2011), and Peregrin (2011). See also Frederick (this volume), and Jespersen (this volume).

subjects at S-structure can be defined in purely syntactic terms. Higginbotham (1987) understands predication as a formal binary relation on points of phrase markers, and Bowers (1993) postulates the existence of a functional category responsible for implementing the relation.[6] Cinque (1992) observes that in the generative approach to linguistic analysis an abstract predication relation underlies all sentences, even those lacking a genuine semantic predication and presentational sentences.

In the Government and Binding model of generative grammar, the obligatory presence of the closing argument (subject) in a subject-predicate structure follows from the second clause of the *Extended Projection Principle* (EPP).[7] Later, Chomsky (1986) suggested that this part of the EPP might be derived from the theory of predication, as developed in Williams (1980) and Rothstein (1985), and observed that the EPP "is a particular way of expressing the general principle that all functions must be saturated" (Chomsky 1986: 116). Chomsky explicitly referred to Frege, and observed that a maximal projection (e.g. VP or AP) may be regarded as a syntactic function that is "unsaturated if not provided with a subject of which it is predicated" (Chomsky 1986: 116).[8]

2. Types of Predication

According to Meixner (2009: 199) "Prior to Frege, there was no philosophically adequate theory of predication (…) Frege, in the nineteenth century, brought the philosophy of predication on the right track".[9] Gottlob Frege has been commonly credited with proposing a bipartite analysis of expressions into a functor and its

---

6 In the semantics developed by Chierchia (1985) and Bowers (1993), the predication operator is a function that takes the property element to form a propositional function, which in turn takes an entity to form a proposition; for further details, see Eide and Åfarli (1999), Åfarli (2005), and Stalmaszczyk (2014).

7 The *Projection Principle* formed in Chomsky (1981: 29) states that the subcategorization properties of each lexical item must be represented categorially at each syntactic level, the EPP adds a second requirement: that clauses have subjects, cf. Chomsky (1982: 10). More recently, Åfarli and Eide (2000) proposed that the *EPP* is the effect of a proposition-forming operation of natural language, induced by a predication operator, and Åfarli (2005) shows the effects of semantic saturation.

8 See the brief discussion in Stalmaszczyk (2014: 243–249). Detailed discussion of predication in contemporary linguistic theory, especially within the generative paradigm, will be provided in the accompanying volume, Stalmaszczyk (ed.) (in preparation).

9 It needs to be stressed, however, that Meixner (2009: 199) immediately adds that Frege's "theory of predication has its own deficits", and offers an appropriate modification of

argument(s). His approach to predication, understood here as a primitive logical relation, may be thus termed *functional*: a function has to be saturated by an argument.[10] Fregean approach contrasts thus with the *concatenative* approach to predication, rooted in the Aristotelian tradition. Both these approaches find application in modern theories – linguistic and philosophical – of language.[11]

2.1 Aristotelian Predication

In Aristotelian semantics, predication is the relation constituted by two elements, the subject with the predicate, with tense specification being a third, concatenating, element, as in the following formula:[12]

Aristotelian predication:
Proposition ⇒ Subject⌢Tense⌢Predicate

Whereas predication in grammatical theory is concerned with linguistic items, Aristotle was concerned with relations between entities in the ontology, it is

---

Frege's theory, "helpful for belief foundation and truth explanation" (Meixner 2009: 213).

10 Fregean predication should be analyzed in connection with his theory of truth and semantic relations. The mutual relations between the notions of truth, existence, identity and predication are discussed in Klement (2002) and Mendelsohn (2005). For a different approach to Fregean predication, in the context of the unity of the proposition, see Gaskin (2008) and Collins (2011). Cf. also, though within a different theoretical context, Sellars (1985) on truth as an essential problem of predication, and Searle (1969) on predication, truth and speech acts. A formal approach to Fregean semantics is presented by Chierchia (1985) and Bowers (1993).

11 Cf. the contributions to Part Two of this volume and to Stalmaszczyk (ed.) (in preparation). This section is based on Stalmaszczyk (2014), where the appropriate terminology (i.e. *functional* and *concatenative* predication) is introduced and justified. For a recent comparison of Aristotelian and Fregean predication (with a discussion of Plato and Leibnitz), see Meixner (2009).

12 Cf. the Aristotelian definition of simple proposition in note 1 above. This is not to claim, however, that Aristotelian predication is limited to structural configurations, on the contrary, it has deep ontological grounding. As observed by Moravcsik (1967: 82), Aristotle "takes predication to be showing the ontological dependence of the entity denoted by the predicate on the entity denoted by the subject". For a different approach to ontological predication in Aristotle, especially its mereological interpretation, see Meixner (2009). For a comprehensive discussion of Aristotelian predication, and especially of its relevance for contemporary philosophy, see Bäck (2010).

therefore necessary to distinguish between reference to linguistic items, i.e. *linguistic predication*, and reference to items in the ontology, i.e. *ontological predication*:[13]

Linguistic predication:
A predicate (a linguistic item) is *linguistically* predicated of its subject (a grammatical item).

Ontological predication:
A predicable (a metaphysical item) is *ontologically* predicated of its subject (an item in the ontology).

Two relations which are variants of metaphysical/ontological predication are introduced by Aristotle in the *Categories*:[14]

(1) X is SAID OF Y
  i. Socrates is a man
  ii. If 'Socrates is a man' is true, then man is SAID OF SOCRATES.

(2) X is IN Y
  i. Socrates is pale
  ii. If 'Socrates is pale' is true, then pallor is IN Socrates.

(1) and (2) exemplify the relations holding between items in the ontology: between a metaphysical subject, Socrates, and a predicable, man or pallor. 'SAID OF a subject' is a relation of ontological classification, 'IN a subject' is a relation of ontological dependence. Furthermore, (1i) tells us something fundamental about what kind of thing Socrates is, it is therefore an example of *essential* predication. On the other hand, (2i) tells us something that happens to be the case, it is an example of *accidental* predication. In (1ii) and (2ii) the linguistic predication is related to ontological predication, however, Aristotle is concerned primarily with giving the metaphysical configurations that underlie sentences (1i) and (2i), and, as pointed out by Lewis (1991: 55), the philosopher is silent on how the two kinds of predication are related. As observed by Lewis (1991: 4, n.4), the relation between linguistic predication and metaphysical predication is not bidirectional: the subject of a linguistic predication can be either a linguistic item or an entity in the ontology, however the subject of a metaphysical predication will always be an ontological item, and not a linguistic one.

A similar point is made by Mesquita (this volume), who stresses the necessity to distinguish two levels in Aristotelian thought: "the ontological level, where

---

13 Lewis (1991: 4) refers to the latter as *metaphysical predication*.
14 Cf. the full discussion in Lewis (1991: 53–63).

we speak of predicates as something that pertains *to* things; and the logical level, where we speak of predicates as something that is said *of* things". In the former case the predicate is an entity 'utterly extra-logical and extra-linguistic', in the latter the predicate is a term, a part of a sentence, hence a linguistic item.

2.2 Fregean Predication

In Fregean semantics the application of a function to an argument is not a mere juxtaposition of the two elements. The function combines with the argument into a self-contained whole due to the fact that it contains a logical gap (the place-holder, or argument-place) which needs filling. As concluded by Frege in "Über Begriff und Gegenstand" ('On Concept and Object', 1892): "not all the parts of a thought can be complete; at least one must be unsaturated or predicative; otherwise they would not hold together" (Frege [1892]: 193). Tichý (1988: 27) comments, somewhat metaphorically, that "the function latches on its argument, sticking to it as if through a suction effect".

Thus, in Fregean semantics, predication is a relation in which an argument saturates an open position in the function, cf. the simplified formula below:

Fregean predication:
Proposition $\Rightarrow$ [Function $(1, ..., n)$]$^\frown$Argument$_{(1, ..., n)}$

It needs to be stressed at this point that Fregean semantics is not concerned with natural language predicates. His line of reasoning, however, may be applied to analyzing predication as a grammatical relation. The above formula aims at capturing Tichý's observation that "the function latches on its argument", furthermore, the "suction effect" is attributed to the presence of the open position(s) – '$(1, ..., n)$'. I use above the term "Proposition" as a generalized term for Frege's "act of judgement" and "assertion".[15] As observed by Stevens:

> Rather than analyzing the proposition into a series of elements (subject, predicate, copula), Frege construes the predicative part of the proposition as a function which is essentially incomplete or 'unsaturated'. (Stevens 2003: 224)

In other words, there is a need for a predicative constituent in a proposition to bind the propositional content into a 'fully-fledged unit', cf. Stevens (2003: 230–231).

---

15 Cf. the distinction introduced by Frege in "Thought" (Frege [1918]: 329):
   (1) The grasp of a thought – thinking,
   (2) The acknowledgement of the truth of a thought – the act of judgment,
   (3) The manifestation of this judgment – assertion.

Paul Pietroski remarked that Frege "bequeathed to us some tools – originally designed for the study of logic and arithmetic – that can be used in constructing theories of meaning for natural languages" (Pietroski 2004: 29–30). These 'Fregean tools' still prove useful in analyzing not only the fundamental issues of sense and reference, but also such notions as predication, predicate, argument and concept, as demonstrated by the chapters in Part Two of this volume, and Frege's "philosophy of language […] remains intensely vital today. Not since medieval times has the connection between logic and language been so close" (Mendelsohn 2005: xviii).

Contributions gathered in this volume discuss the philosophy and logic of predication, they focus on the Aristotelian model and its legacy, and also on Fregean predication, its limits and possible refinements.

3. Contents of the Volume

The volume is divided into two parts: papers in Part One are concerned with historical investigations, albeit with considerable importance for contemporary research on predication theory and implications for the future. Contributions in Part Two discuss philosophical and more formal topics, including an algebraic approach to Meinongian predication, and tackle some controversial issues stemming from Fregean semantics.

Predication is a complex entity in Aristotelian thought. António Pedro Mesquita attempts to provide an account for this complexity, starting with a crucial chapter of the *Posterior Analytics* (I 22), where, in the most complete and developed manner within the corpus, Aristotle proceeds to systematize this topic. It follows from the analysis that predication can assume, generically, five forms: the predication of essence (that is of the genus and the specific difference); essential predication (that is either of the genus or of the differences or their genera); the predication of accidents per se; the predication of simple accidents; and accidental predication. Mesquita observes that only some types are forms of strict predication (ἁπλῶς), and he concludes with a discussion of Aristotle's thesis according to which no substance can be a predicate, which is implied by its notion of accidental predication, a thesis which has been (wrongly so in the author's opinion) challenged in modern times.

Nino B. Cocchiarella offers a conceptualist view on predication. He starts with observing that predication can be also understood as the nexus between a subject and a predicate expression, the basis of the unity of a speech act, including speech acts in the plural and speech acts that involve mass nouns. And a speech act, of course, is an overt expression of a mental act, e.g., a judgment; and therefore the unity of a speech act such as an assertion is really the unity of the judgment that underlies that act. Such a mental act, and therefore the speech act as well, has a unity based on how the referential and predicable roles of the subject and predicate expressions combine and function together respectively. Cocchiarella proposes to explain this unity of predication in terms of a conceptualist theory of logical forms that, as he claims, underlies at least some important aspects of thought and natural language. Cocchiarella's conceptualist logic also contains an account of the medieval identity (two-name) theory of the copula, as well as an account of plural and mass noun reference and predication, the truth conditions of which are based on a logic of plurals and mass nouns.

Ignacio Angelelli briefly discusses the impact of traditional predication theory on the notion of class. He quotes a text from the 18th c. by Johann Caspar Sulzer, where not only the notion of class (*classis*) is used but, moreover, a hierarchy of orders for classes is introduced. Such a hierarchy is developed not along the relation of membership but along the relation of inclusion. Given a class of order n, a class of order $n+1$ is not one which has the class of order n as a member but one of which the class of order n is a subclass. The main thesis of Angelelli's paper is that the ordering of classes by inclusion rather than by membership is a consequence of the traditional, as opposed to the Fregean, predication theory.

Allan Bäck looks at the future of predication theory. He observes that in the history of predication, an emphasis has been placed on impersonal statements, a move from speaker meaning to sentence meaning, an elimination of intentionality and standpoint via the elimination of indexicals, and a move from ordinary to ideal language. Following these trends in light of their historical success, Bäck suggests that a theory of predication with the following features should be developed: formalism, having a sharp distinction between the pure theory and its interpretations, and the primacy of two-place relations. Accordingly, he offers a theory of predication with the following features: an uninterpreted language, based on the combinatorials of primitive symbols and on the application of formation rules so as to give well-formed formulae, construction of n-place predicates via unsaturating the wffs, the reduction of all n-place predicates to two-place predicates (relations). Bäck concludes with considering how successfully such a predication theory might be interpreted and applied, in particular to metaphysics.

Contributions in Part Two discuss some controversial issues in Fregean semantics, and they offer different approaches to contemporary predication theory.

Mieszko Tałasiewicz examines the notion of the sense of a predicate, as stemming from Frege, in the light of the controversy between Michael Dummett and Peter Geach. This controversy is concerned with inquiring as to whether the sense of a predicate is a way of getting to the reference of the predicate or rather a function from senses of names to senses of sentences. Several Fregean principles are discussed as well as the distinction of simple and complex predicates. Tałasiewicz concludes that Functoriality Principle can be simultaneously applied with Principle of Mediating Sense, so that the surface form of the Dummett-Geach controversy dissolves. However, Compositionality Principle cannot be held simultaneously with Functoriality Principle, so that there arises a question which principle is worth keeping in the first place. The paper argues that Functoriality Principle should stay, because there is no other sensible way to obtain what it gives (namely, the unity of the proposition); while there are alternative ways (involving situation semantics) of explaining facts that in Fregean framework were explained through Compositionality Principle.

Danny Frederick is concerned with Frege's theory of unsaturatedness. He observes that Frege proposed his doctrine of unsaturatedness as a solution to the problems of the unity of the proposition and the unity of the sentence. In his contribution Frederick shows that Frege's theory fails, and that it entails that singular terms cannot be predicates. Frederick explains the traditional solution to the problem of the unity of the sentence, as expounded by John Stuart Mill, which invokes a syncategorematic sign of predication and the connotation and denotation of terms. He contrasts the resulting conventionalist account with Frege's unsaturatedness account and argues that the conventionalist account provides a clear and intelligible solution to the problem of the unity of the sentence which is free of the defects of Frege's account. Finally, he suggests that the problem of the unity of the proposition is spurious, and recommends that the notion of unsaturatedness be excluded from serious debate.

David Liebesman discusses yet another issue, of constant relevance, deriving from Frege's semantics: the concept *horse*. He observes that far from being of mere historical interest, concept horse-style expressibility problems arise for versions of type-theoretic semantics in the tradition of Montague. Grappling with expressibility problems yields lessons about the philosophical interpretation and empirical limits of such type-theories.

Peter Hanks focuses on predication and rule-following. He observes that according to the traditional view, propositions are the primary bearers of truth conditions. We latch onto propositions by entertaining them, and then deploy them in various ways in our thoughts and utterances. The truth conditions of our thoughts and utterances are then derived from the truth conditions of propositions. This raises the question of whether we can explain how propositions themselves have truth conditions, which is one form of the problem of the unity of the proposition. Attempts at solving this problem end in a philosophically unsatisfying account of representation in thought and language, in which propositions and our relations to them are all taken as primitive and *sui generis*. Hanks discusses Platonism about rule-following, the metaphysical and epistemological problems for Platonism, and attempts to clarify the problems that Wittgenstein's rule-following considerations pose for an act-based theory of predication; and finally, he presents some desiderata for a solution.

Bjørn Jespersen observes that predication may be conceived as an *act* that speakers or thinkers execute when predicating a property of an individual, or as an *operation* that applies to a pair consisting of a property and an individual. In his contribution, Jespersen contrasts a logical or objectivist conception of predication with a pragmatic or subjectivist conception. The former one identifies predication with an objective operation that applies a property to an individual, whereas the latter conception identifies predication with an act that a speaker or thinker carries out when asserting or thinking that an object has a property. Jespersen argues that the two conceptions of predication are complementary rather than mutually exclusive, but also that the logical (operation-theoretic) conception enjoys conceptual priority over its act-theoretic counterpart.

Finally, Jacek Paśniczek discusses Meinongian predication. Alexius Meinong, an Austrian philosopher, is particularly famous for his abundant ontology embracing various types of non-existent objects. The most distinguished formal-ontological feature of his theory of objects is that it allows for impossible and incomplete objects like *the round square*. Recently there have appeared several logical interpretations of Meinong's theory of objects rendering it consistent. In order to render such a theory consistent, the classical concept of predication according to which it is a relation holding between objects and properties (i.e. basic formulas of respective formal languages are subject-predicate ones) has to be modified. Paśniczek develops an unorthodox view of Meinongian predication based on a calculus of names and provides an appropriate algebraic approach to predication.

Whereas contributions gathered in this volume discuss the philosophy and logic of predication, an accompanying volume, Stalmaszczyk (ed.) (in preparation), will present approaches to predication theory in contemporary linguistics, especially within the generative paradigm.

I wish to thank all the authors involved in this enterprise for submitting their papers, and also for their willingness to peer review, and to discuss the organization of the volume.

References

Åfarli, T. (2005). "Predication in Syntax". *Research in Language* 3: 67–89.

Åfarli, T. and Eide, K. (2000). "Subject Requirement and Predication". *Nordic Journal of Linguistics* 23: 27–48.

Aristotle [1941]. *De Interpretatione* (On Interpretation). Translated by E. M. Edghill. In: *The Basic Works of Aristotle*. Edited by R. McKeon. New York: Random House, 38–61.

Bäck, A. (2010). *Aristotle's Theory of Predication*. Leiden-Boston-Köln: Brill.

Beaney, M. (ed.) (1997). *The Frege Reader*. Oxford: Blackwell Publishers.

Bloomfield, L. (1933). *Language*. New York: Holt [Reprinted London: George Allen & Unwin, 1967].

Bowers, J. (1993). "The Syntax of Predication". *Linguistic Inquiry* 24: 591–656.

Chierchia, G. (1985). "Formal Semantics and the Grammar of Predication". *Linguistic Inquiry* 16: 417–443.

Chomsky, N. (1981). *Lectures on Government and Binding*. Dordrecht: Foris.

Chomsky, N. (1982). *Some Concepts and Consequences of the Theory of Government and Binding*. Cambridge, MA: MIT Press.

Chomsky, N. (1986). *Knowledge of Language. Its Nature, Origin, and Use*. New York: Praeger.

Cinque, G. (1992). "Predication". In: W. Bright (ed.), *International Encyclopedia of Linguistics, Volume 3*, New York and Oxford: Oxford University Press, 368–269.

Cocchiarella, N. B. (1989). "Philosophical Perspectives on Formal Theories of Predication". In: D. Gabbay and F. Guenthner (eds.), *Handbook of Philosophical Logic, Volume 4*, Dordrecht: D. Reidel Publishing Co., 253–326.

Cocchiarella, N. B. (2013). "Predication in Conceptual Realism". *Axiomathes* 23 (2): 301–321. doi:10.1007/s10516-010-9140-x.

Collins, J. (2011). *The Unity of Linguistic Meaning*. Oxford: Oxford University Press.

Davidson, D. (1967). "The Logical Form of Action Sentences". [Reprinted in D. Davidson, 1980. *Essays on Actions and Events*, Oxford: Clarendon Press, 105–122.]

Davidson, D. 2005. *Truth and Predication*. Cambridge, MA and London: Belknap Press of Harvard University Press.

Dudman, V. H. (1985). "Towards a Theory of Predication for English". *Australian Journal of Linguistics* 5: 143–196.

Eide, K. and Åfarli, T. (1999). "The Syntactic Disguises of the Predication Operator". *Studia Linguistica* 53: 155–181.

Frege, G. [1892] (1997). "On Concept and Object" (translated by P. Geach). In: M. Beaney (ed.), 181–193.

Frege, G. [1918] (1997). "Thought" (translated by P. Geach and R. H. Stoothoff). In: M. Beaney (ed.), 325–345.

Gaskin, R. (2008). *The Unity of the Proposition*. Oxford and New York: Oxford University Press.

García-Carpintero, M. (2010). "Gaskin's Ideal Unity". *Dialectica* 64: 279–288.

Higginbotham, J. (1987). "Indefiniteness and Predication". In: E. Reuland and A. ter Meulen (eds.), *The Syntactic Forms of Predication*. Bloomington: Indiana University Linguistics Club, 43–70.

Hockett, C. F. (1958). *A Course in Modern Linguistics*. New York: The Macmillan Company.

Jespersen, O. (1937). *Analytic Syntax*. Helsingor. [Reprinted New York: Holt, Rinehart and Winston, 1969].

Klement, K. C. (2002). *Frege and the Logic of Sense and Reference*. London and New York: Routledge.

Krifka, M. (1998). "The Origins of Telicity". In: S. Rothstein (ed.), *Events and Grammar*. Dordrecht: Kluwer Academic Publishers, 197–235.

Lewis, F. A. (1991). *Substance and Predication in Aristotle*. Cambridge: Cambridge University Press.

Link, G. (1998). *Algebraic Semantics in Language and Philosophy*. CSLI Lecture Notes No. 74, Stanford: Center for the Study of Language and Information.

Lorenzen, P. (1968). *Methodisches Denken*. Frankfurt am Main: Suhrkamp Verlag.

Meixner, U. (2009). "From Plato to Frege: Paradigms of Predication in the History of Ideas". *Metaphysica. International Journal for Ontology and Metaphysics* 10: 199–214.

Mendelsohn, R. (2005). *The Philosophy of Gottlob Frege*. Cambridge: Cambridge University Press.

Moravcsik, J. M. E. (1967). "Aristotle on Predication". *Philosophical Review* 76: 80–96.

Peregrin, J. (2011). "There is no such Thing as Predication". *Conceptus* 97: 29–51

Pietroski, P. M. (2004). *Events and Semantic Architecture*. Oxford: Oxford University Press.

Quine, W. V. O. (1960). *Word and Object*. Cambridge, MA: MIT Press.

Reichenbach, H. (1947). *Elements of Symbolic Logic*. Berkeley: University of California Press. [Reprinted New York: The Free Press, 1966].

Rothstein, S. D. (1985). *The Syntactic Forms of Predication*. Bloomington: Indiana University Linguistics Club.

Rothstein, S. D. (1992). "Predication and the Structure of Clauses". *Belgian Journal of Linguistics* 7: 153–169.

Searle, J. (1969). *Speech Acts: An Essay in the Philosophy of Language*. Cambridge: Cambridge University Press.

Sellars, W. (1985). "Towards a Theory of Predication". In: J. Bogen and J. McGuire (eds.), *How Things Are*, Dordrecht: D. Reidel Publishing Co., 285–322.

Stalmaszczyk, P. (2014). "The Legacy of Frege and the Linguistic Theory of Predication". In: P. Stalmaszczyk (ed.), *Philosophy of Language and Linguistics. The Legacy of Language and Linguistics*. Berlin and Boston: De Gruyter, 225–253.

Stalmaszczyk, P. (ed.) (in preparation). *Linguistic Theories of Predication*. Frankfurt am Main: Peter Lang.

Stevens, G. (2003). "The Truth and Nothing but the Truth, yet Never the Whole Truth: Frege, Russell and the Analysis of Unities". *History and Philosophy of Logic* 24: 221–240.

Strawson, Peter F. (1971). *Logico-Linguistic Papers*. London: Methuen.

Tichý, P. (1988). *The Foundations of Frege's Logic*. Berlin and New York: Walter de Gruyter.

Williams, E. (1980). "Predication". *Linguistic Inquiry* 11: 203–238.

Part I.
From Aristotle to the Future

António Pedro Mesquita
University of Lisbon

Aristotle on Predication[1]

Abstract: Predication is a complex entity in Aristotelian thought. The aim of the present essay is to account for this complexity, making explicit the diverse forms it assumes. To this end, we turn to a crucial chapter of the *Posterior Analytics* (I 22), where, in the most complete and developed manner within the *corpus*, Aristotle proceeds to systematize this topic.

From the analysis, it will become apparent that predication can assume, generically, five forms: 1) the predication of essence (τὸ αὐτῷ εἶναι κατηγορεῖσθαι), that is of the genus and the specific difference; 2) essential predication (τό ἐν τῷ τί ἐστι κατηγορεῖσθαι), that is either of the genus or of the differences (or their genera); 3) the predication of accidents *per se* and 4) simple accidents (ὡς συμβεβηκότα κατηγορεῖσθαι); 5) accidental predication (κατὰ συμβεβηκὸς κατηγορεῖσθαι).

However, only types 2–4 are forms of strict predication (ἁπλῶς). In effect, the "predication" of essence is not a genuine predication, but a formula for identity, constituting, technically, the statement of the essence of the subject (or its definition). On the other hand, accidental "predication" can only be conceived of as such equivocally, since it results from a linguistic accident through which the ontological subject of the attribution suffers a displacement to the syntactic position of the predicate, which is not, by nature, its own. In neither case does the attribution bring about any legitimate predication.

The study concludes with a discussion of Aristotle's thesis according to which no substance can be a predicate, which is implied by its notion of accidental predication, a thesis which has been – and in our opinion wrongly so – challenged in modern times.

Keywords: Aristotle, *Posterior Analytics*, predication, predication of essence, essential predication, predication of accidents, accidental predication.

0. Introduction

Predication is a complex entity in Aristotle's thought. The object of the present paper is to account for that complexity, rendering explicit the several forms it assumes.

Given the significance of this concept in Aristotle's logic and ontology, the task is relevant *per se*. It is, however, particularly important to avoid the confusion that can easily set in between two concept pairs whose members Aristotle is careful

---

1 An earlier version of this paper was published, under the title "Types of Predication in Aristotle (*Posterior Analytics* I 22)", in: *Journal of Ancient Philosophy*, 2012, 6: 1–27.

to discriminate: one, accidental predication (κατὰ συμβεβηκὸς κατηγορεῖσθαι) as different from predication of accidents (ὡς συμβεβηκότα κατηγορεῖσθαι); the other, essential predication (τό ἐν τῷ τί ἐστι κατηγορεῖσθαι) as different from predication of the essence (τὸ αὐτῷ εἶναι κατηγορεῖσθαι). And this is so because, for Aristotle, neither accidental predication nor predication of the essence is, strictly speaking, *predication*, but rather the "lower" and "upper" margins within whose scope predication is defined.

The "upper" limit – predication of the essence – is definition. The distinction between definition and approximate forms of predication (viz., essential predication) is crucial to set up a precise distinction between predication and definition and to understand the singularity the notion of definition holds within the set of attributive statements in Aristotle.

The "lower" limit corresponds to that which Philoponus dubbed "counternatural predication" (παρὰ φύσιν), so as to distinguish it from predication proper or, as he would call it, "natural predication" (κατὰ φύσιν)[2] – clearly, a heavily symbolic classification.

It is in a crucial chapter of the *Posterior Analytics* (I 22), a chapter which apparent purpose is merely to show the impossibility of an infinite chain of premises in demonstration, that we can find, in a thorough and systematic manner, Aristotle's schematisation of the various types of predication.

Accordingly, it will be by addressing this chapter, in the form of a running commentary on each of its significant units, that we will attempt to follow Aristotle's lesson on this issue.

In the end, we will draw some consequences regarding a strong thesis of Aristotle's theory of predication, viz., that no individual can be a predicate.

1. Strict Predication and Accidental Predication

1.1 Text[3]

> In the case of predicates constituting the essential nature of a thing [τῶν ἐν τῷ τί ἐστι κατηγορουμένων], the situation is clear: if definition is possible, or, in other words, if essential form is knowable,[4] and an infinite series cannot be traversed, predicates

---

2   Cf. *In APo.* 236.24–26 Wallies.
3   *APo.* I 22, 82b37–83a17. (All translations of this chapter are Mure's, with corrections.)
4   Here, the conjunction ἤ clearly holds epexegetic, not disjunctive, value (thus Mure, Tredennick, Tricot; Barnes, Pellegrin).

constituting a thing's essential nature [τὰ ἐν τῷ τί ἐστι κατηγορούμενα] must be finite in number.

But as regards predicates generally we have the following prefatory remarks to make. We can affirm without falsehood that the white (thing) is walking and that that big (thing) is a log; or again, that the log is big and that the man walks. But the affirmation differs in the two cases. When I affirm that the white is a log, I mean that something which happens to be white is a log [ὅτι ᾧ συμβέβηκε λευκῷ εἶναι ξύλον ἐστίν], not that white is the subject in which log inheres; for it is not because it is white or precisely a certain type of white [οὐδ' ὅπερ λευκόν τι] that the white (thing) comes to be a log. Therefore, the white (thing) is not a log except by accident [ὥστ' οὐκ ἔστιν ἀλλ' ἢ κατὰ συμβεβηκός]. On the other hand, when I affirm that the log is white, I do not mean that something else, which happens also to be a log [ἐκείνῳ δὲ συμβέβηκε ξύλῳ εἶναι], is white (as I should if I said that the musician is white, which would mean that the man who happens also to be a musician [ᾧ συμβέβηκεν εἶναι μουσικῷ] is white); on the contrary, log is here the subject, which actually came to be white and did so because it is a log or precisely a certain log, not because it is something else. If, then, we must lay down a rule, let us entitle the latter kind of statement predication [κατηγορεῖν], and the former not predication at all, or not strict [ἁπλῶς] but accidental predication [κατὰ συμβεβηκὸς].

1.2 Comment

In these two paragraphs, Aristotle drafts a preliminary enumeration of several types of predication: essential predication;[5] predication proper, or strict predication;[6] and accidental predication.[7]

Strict predication (ἁπλῶς) and accidental predication (κατὰ συμβεβηκός) are clearly distinguished at the end of the passage as opposite types of predication.

The text is, at this point, particularly interesting.

The distinction between accidental and strict predication is there made to depend on a metaphysical interpretation of the subject/predicate pair, namely, that not every term that can fill the predicate's logical or syntactic slot in a sentence refers to a predicate in the ontological sense, and particularly the actual predicate of the thing referred to by the sentence's subject, that is, a property that actually belongs to it. A more basic distinction is here being assumed between that which is a predicate by nature, i.e., that *which is said of something* (of a "natural" subject), and that which is a subject by nature, i.e., that *of which something* (a "natural" predicate) *is said*.

---

5   82b37–83a1.
6   83a9–14.
7   83a4–9.

Aristotle's thesis that no individual (or, in the terms of the *Categories*, no primary substance) can be a predicate is here justified. What it states is that every individual is "naturally" a subject, for which reason it cannot be (from an ontological point a view) a predicate. When an individual comes to be a predicate (from a logical or syntactic point of view), which is to say, in more rigorous terms, when it happens that the name of an individual, or, in general, a singular term, fills the predicative slot in an attributive sentence, this happens in a merely accidental way, i.e., by virtue of a linguistic accident that abusively shifts it to that inappropriate slot.[8]

Now, this is the assumption that justifies the distinction between strict predication (where subject and predicate are "natural") and accidental predication, where subject (e.g., musician in "The musician is white") or both subject and predicate (e.g., "That white thing is a log") are not "natural".[9]

We can thus say that, concerning the distinction between these two types of predication, the late Neoplatonic nomenclature that dubbed them "natural" and "counternatural", respectively, albeit not introduced by the Stagirite, quite aptly reflects the spirit of his doctrine in this regard.

It is worth pointing out that the relation of either one or both types of predication to the essential predication mentioned in the first paragraph is nowhere clarified. Furthermore, it is not explicit whether such predication should be included under strict predication or, on the contrary, whether it should be understood as some autonomous type to which the two other types of predication distinguished in the second paragraph would jointly oppose.

In this circumstance, the table resulting from the two initial paragraphs can be, quite simply, as follows:

---

8 See, typically, *APr.* I 27, 43a32–36: "It is clear then that certain things are not naturally said of anything [ἔνια τῶν ὄντων κατ' οὐδενὸς πέφυκε λέγεσθαι]: in fact, each sensible thing has such a nature that it cannot be predicated of anything, *save by accident* [πλὴν ὡς κατὰ συμβεβηκός], as when we say that that white thing is Socrates [τὸ λευκὸν ἐκεῖνο Σωκράτην εἶναι] or that that thing that approaches us is Callias [καὶ τὸ προσιὸν Καλλίαν]."
9 Along the same lines, cf. *Metaph.* Δ 7, 1017a7–22, and also: *Int.* 11, 21a7–16; *APr.* I 27, 43a32–43; *APo.* I 4, 73b5–10; *APo.* I 19, 81b23–29. Other occurrences in: *APo.* I 13, 79a6; *Ph.* I 4, 188a8; *Metaph.* A 6, 987b23; B 4, 1001a6; B 4, 1001a10; B 4, 1001a 28; N 1, 1087a33; N 1, 1087a 35; N 1, 1088a28.

Aristotle on Predication

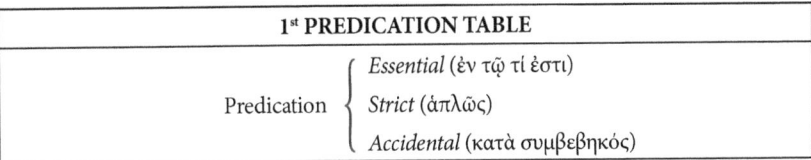

1st PREDICATION TABLE

Predication { *Essential* (ἐν τῷ τί ἐστι) / *Strict* (ἁπλῶς) / *Accidental* (κατὰ συμβεβηκός) }

However, given that all examples added in the second paragraph are examples of non-essential predication, one could assume that the distinction Aristotle introduced therein between strict predication and accidental predication is not to be added to the type mentioned in the first paragraph, but to oppose to it, which would entail reformulating the table thus:

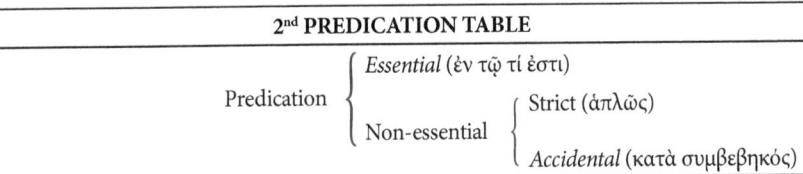

2nd PREDICATION TABLE

Predication { *Essential* (ἐν τῷ τί ἐστι) / Non-essential { *Strict* (ἁπλῶς) / *Accidental* (κατὰ συμβεβηκός) } }

This is probably why the differentiation between strict and accidental predications opens the subsequent text, which is aimed at excluding the latter, but not the former, from the discussion.

2. Strict Predication

2.1 Text[10]

White and log will thus serve as types respectively of predicate and subject. We shall assume, then, that the predicate is invariably predicated strictly [ἁπλῶς] and not accidentally [ἀλλὰ μὴ κατὰ συμβεβηκός] of the subject, for on such predication demonstrations depend for their force. It follows from this that when a single attribute is predicated of a single subject, the predicate must affirm of the subject either some element constituting its essential nature [ἢ ἐν τῷ τί ἐστιν], or that it is in some way qualified, quantified, related, active, passive, placed, or dated.

---

10 *APo.* I 22, 83a17–23.

2.2 Comment

The content of the present paragraph can be captured in the following theses:

1) In every predicative sentence, a predicate stands in the same relation to the subject as 'B' stands to 'A' in the standard sentence 'A is B'.
2) The predicate can be predicated of the subject either strictly or accidentally.
3) In canonical, or strict, predicative sentences, the predicate stands in the same relation to the subject as "white" stands to "log" in the sentence "The log is white". (Up to this point, we have merely summed up the doctrine accounted for in the previous paragraph.)
4) Now, every strict predication abides by the table of categories; therefore, in such predication, the predicate says of the subject either what the subject is, or of which type it is, or in relation to what it is, etc.
5) In the first of the mentioned cases in (4), the predication is essential predication (ἢ ἐν τῷ τί ἐστιν).
6) In all remaining cases, it will certainly be strict, but not essential, predication.

The consequences of this clarification for our subject matter, particularly for solving the problem left suspended in section 1, are evident.

Following this clarification, strict predication is the predication type that can be essential or non-essential, in which case the former is rehabilitated (and given the same status as the latter) as a type of strict predication.

We may now use the data from the current paragraph to put forward a third predication table:

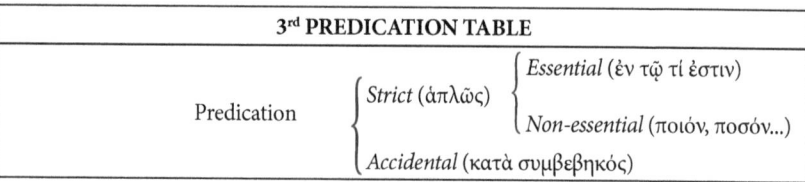

There is a staggering difference between the second and third tables:

2nd PREDICATION TABLE	3rd PREDICATION TABLE
Predication { Essential / Non-essential { Strict / Accidental } }	Predication { Strict { Essential / Non-essential } / Accidental }

In the left-hand side table, essential predication is opposed to both strict predication and accidental predication, which are there presented as two types of non-essential predication.

In the right-hand side table, essential predication is placed under strict predication, and it is the latter that, in both its variants (essential predication and non-essential predication), is now opposed to non-strict, or accidental, predication.

Clearly, the point of contrast concerns which locus to attribute to essential predication.

Before such a huge discrepancy between the two classifications, to which should we ascribe greater value? Should essential predication be considered a sub-type of strict predication (third table) or, on the contrary, it is to stand as a predication type, side-by-side with non-essential (strict and accidental) predication (second table)?

This can only be decided in light of the information provided in the subsequent paragraphs.

3. Strict Predication *(a)* of the Substance and *(b)* of Accidents

3.1 Text[11]

Predicates which signify substance signify precisely the subject, or a certain type of the subject. Predicates not signifying substance which are predicated of a subject which is neither precisely what the subject is, nor a certain type of what the subject is, are accidents [συμβεβηκότα]. For instance, when you predicate white of man, man is not precisely white or precisely a certain type of white [οὐ γάρ ἐστιν ὁ ἄνθρωπος οὔτε ὅπερ λευκὸν οὔτε ὅπερ λευκόν τι], but rather animal, since man is precisely an animal [ὅπερ γὰρ ζῷόν ἐστιν ὁ ἄνθρωπος]. These predicates which do not signify substance must be predicates of some other subject, and nothing can be white which is not white because of something else.

3.2 Comment

In this paragraph, Aristotle sets forth a double alternative to what we have so far been indistinctly calling "essential predication".

On the one hand, some essential predicates (or "things which signify substance") are those that "signify *precisely* the subject". In this case, essential predicates express the nature or identity of the subject itself, and therefore, in the predicative sentence, the predicate is identical to the subject.[12]

---

11 *APo.* I 22, 83a24–32.
12 The terms "identical", "identification", "identity", which we will henceforth systematically use, always possesses intensional value, expressing the interchangeability of the

On the other hand, essential predicates can also be those that signify "*a certain type* of what the subject is". In this instance, essential predicates do not express the subject itself, but that under which the subject falls in the generative scale, namely, a genus or a differentia of the subject.[13]

Considering the alternative here proposed, we may now understand that essential predication *in a certain sense is* and *in a certain sense is not* strict predication, which allows us to understand the reason for the discrepancy between the second and third tables, as well as to solve it, bringing it to a more enlightening compatibility.

In a certain sense it is, and in another sense it is not, strict predication, because it itself already has two meanings, namely, those two we have just introduced.

Let us be quite clear, though, as to what this means. It is not that, in Aristotle, the present sentence on the one hand excludes, while on the other hand includes, essential predication from strict predication. In fact, from Aristotle's point of view, the characterisation of essential predication here introduced is stated *against non-essential predication* and *within the general framework of strict predication*. This much is shown in the fact that Aristotle proceeds to this characterisation after having restricted the discussion to strict predication,[14] and by his introducing in the next lines, as a contrast, predication of accidents[15] as a second type of strict predication. Accordingly, essential predication is here presented as a sub-type of that which is called, simply, "to predicate" (κατηγορεῖν ἁπλῶς). And this is clearly coherent with the fact that predication ἁπλῶς abides by the table of categories, wherein the substance (under which essential predicates fall) is merely a category amongst others.

For Aristotle, the question is therefore simple: either there is real predication, in which case it abides by the table of categories, where essential predicates are included; or there is no predication at all, unless in a certain accidental sense.

When we limit ourselves to predication ἁπλῶς, as Aristotle does in the beginning, essential predication comes to be but a kind of strict predication, or, simply put, one kind of predication.

---

    subject and the predicate and not just their simple co-extensionality. Technically speaking, in Aristotle, the latter constitutes predication proper (ἴδιος), whilst the former is the definition (cf. *Topics* I 4–5, 8).

13 Cf. 83b1.
14 83a17–23.
15 83a25–35.

In this light, to technically distinguish predication ἐν τῷ τί ἐστι, predication ἁπλῶς and predication κατὰ συμβεβηκός, as we did in our first table, is to distance ourselves from the Aristotelian classification.

This is so because, for Aristotle, either there is or there is not predication. If there is, then predication can *as well* be essential (ἐν τῷ τί ἐστι). If there is not, it can nevertheless occur *accidentally* (κατὰ συμβεβηκός).

In Aristotle's view, the crucial divide stands thus between (strict) predication and accidental predication. In face of this divide, essential predication has virtually no specificity at all (except, of course, to the extent that it is one of the types in which predication is subdivided).

All this appears to definitively establish the third predication table as the correct one from an Aristotelian perspective. However, this is not so. And it is not so *precisely because of the passage we are currently commenting*. Despite what Aristotle could have (or would have liked to have) expressly acknowledged, this passage sets the grounds for a new account of strict predication, one which opposes not just accidental predication, but also essential predication, or, at least, a certain type of essential predication. It is, therefore, essential predication itself which is, in this clause, implicitly reassessed. This reassessment enables us to understand why essential predication in a sense is, and in another sense is not, strict predication and, above all, it enables us to understand *in what sense* essential predication is, and *in what sense* it is not, strict predication.

From a general point of view, essential predication is, as already seen, strict predication, for predication ἐν τῷ τί ἐστι predicates under the category of substance. However, the adjective "strict" means here only that essential predication is *simply* (ἁπλῶς) a type of predication and not a kind of pseudo-predication – a predication "by accident".

That said, if we pay close attention to the nature of essential predication, as it is here defined by Aristotle, we realise that there is something fundamental that sets it apart from every other type of predication ἁπλῶς. In this sense, the term ἁπλῶς acquires a new meaning, circumscribing everything that is predication proper, as opposed to accidental "predications" which, *due to some motive*, are not genuine predications, but also as opposed to essential "predications" which, *due to another motive*, are not, likewise, genuine predications.

Accidental "predications" are not genuine predications because the sentence's predicate does not refer to an actual property of the thing referred to by the sentence's subject, i.e., something that truly belongs to it. In Aristotle's terms, in sentences expressing such "predications", the predicate is not attributed to the subject in virtue of the subject being precisely what it is, but because something else (sometimes, the predicate itself) is, accidentally, that subject. Thus, in "The musician is white", it is not because the musician is a musician, but because *there is a certain man who happens to be a musician*, that the predicate is (accidentally) attributed to the subject. Likewise, in "that white thing is a log", it is not because that white thing is white, but because *there is a certain log which happens to be white*, that the predicate is (accidentally) attributed to the subject. In this sense, the reason why accidental "predications" are not genuine predications is that one necessary condition of predication is not fulfilled: the predicate is not attributed to the subject because the subject is what it is (or, which is the same, the predicate is not attributed to the subject as something that really belongs to it). In accidental predications, what we see is that, by virtue of a syntactic accident, something that is not a "natural" predicate, or a "natural" subject, shifts, in the sentence, into a logical place that does not "naturally" belong to it.

Now, in the case of essential "predications", this requirement is fulfilled. But, in a way, it is *excessively* fulfilled, for, in this case, the predicate is not simply attributed to the subject *because* the subject is precisely what it is, but because the subject *is precisely that predicate*.

Accordingly, whereas in accidental "predications" the predicative link does not truly exist, for the sentence's predicates do not refer to actual properties of the subject, in essential "predications", the predicative link is not truly predicative, for the sentence's predicates do not refer to properties of the subject *in the strict sense of the word* (ἁπλῶς) – they refer to the subject itself.

In a word, essential "predications" are not, for Aristotle, genuine predications, but identity formulae. They must thus be distinguished from strict predication, just as it happened with accidental "predications", albeit for a different reason.

Granted that nowhere in this chapter does Aristotle expressly draw this conclusion. However, in an overall context, this conclusion is required by the characterisation of predication *qua* attribution "of something to something" (τὶ κατὰ τινός),[16] or "of another to another" (ἕτερον καθ' ἑτέρου),[17] or still "of one to one" (ἓν καθ' ἑνός),[18] whereas the attribution of essence is a process "of the same to the same" (αὐτὰ αὐτῶν).[19]

In sum, essential "predication" cannot be strictly considered as predication, in that it is *a definition*: and a definition does not say something of something, but

---

16 Cf. *Int.* 6, 17a25 (and 3, 16b6–10); *APr.* I 1 24a16.
17 Cf. *Cat.* 3, 1b10.
18 See, especially, *APo.* I 22, 83b17–19.
19 Cf. *ibid.*

simply *the* something;[20] it does not say of something *that [it] is* something, but merely *what* the something is.[21]

It should be noted that this concords with the distinction, consistently assumed by Aristotle, in the context of the classification of the principles of demonstration,[22] between saying "that it is" (ὅτι ἔστι) and saying "what it means" (τί σημαίνει): definitions do not say that something is something, they merely say what something means. Therefore, only axioms and theses (hypotheses and postulates) are predications – not so definitions. Definitions are not so because they do not truly contain a τὶ κατά τινος λέγεσθαι.

Now, this allows us to understand why is it that essential predication *is and is not strict predication* and the sense in which it *is* and the sense in which it *is not* strict predication.

In fact, everything we have developed throughout the present point is valid for definitions only: and what the doctrine introduced in this passage shows exactly is that *not every essential predication is a definition*.

Aristotle distinguishes between essential predicates that mean *precisely that of which they are predicated* and those that mean *a certain type* of that of which they are predicated.

Let us recall an excerpt already cited:[23]

> For instance, when you predicate white of man: man is not precisely white or precisely a certain type of white, but rather animal, since man is precisely an animal.

That is: in the predication "the man is white", subject and predicate are not the same, because the man is not the white, nor a certain type of white (a specific kind of white). But in the predication "man is an animal", subject and predicate are the same, because man is a (certain type of) animal, i.e., a specific kind of animal.

In the former predicative sentence, that which is attributed is, therefore, an accident of man (white), whereas the latter attributes that of which man is a species (animal).

---

20 See *APo.* II 4, 91b1-7 (and cf. 91a15-16; II 6, 92a6-9; II 13, 96a20-b1); *Top.* I 5, 102a13-14 (and cf. VII 2, 152b39-153a1); *Metaph.* Z 4, 1030a7-11.
21 Paradigmatically in *APo.* II 3, 90b38-91a2: "Furthermore, to prove *what it is* [τὸ τί ἐστι] and *that it is* [ὅτι ἔστι] is different. Definition shows what it is, while demonstration [shows] that this is or is not [said] of that [ἡ δὲ ἀπόδειξις ὅτι ἔστι τόδε κατὰ τοῦδε ἢ οὐκ ἔστιν]." But cf. also *APo.* I 1, 71a11-17; 2, 72a18-24; 10, 76b35-77a4.
22 Cf. *APo.* I 1, 71a1-17; 2, 72a14-24; 10, 76a31-36.
23 83a28-30.

Now, in general, these two examples outline the distinction between predication of accidents and predication of the substance as *types of strict predication*. However, if we were to add to them the example Aristotle does not provide in this step, viz., "man is a biped animal", where subject and predicate are the same (for man *is precisely what to be a biped animal is*) the existence of a further type of attributive statement would clearly follow – one that would no longer be strict predication, but instead more-than-strict (so to speak), or hyperbolic, predication, for in it the predicate is *precisely* the subject.

The difference between the two types of essential predication is now clear: in general predication under the category of substance, the subject is not identified *with the* predicate (man is not animal); instead, it is identified *as* "a certain type" of the predicate (man is a certain kind of animal). In predication of the essence, on the contrary, the subject is identified with the predicate itself (man is a biped animal); we have, thus, *a definition*. In other words, the copulative relation is not, in the latter case, from predicate to subject, rather from *definiens* to *definiendum*.

That is why the distinction between the two types of sentence is, from a logical point of view, quite clear too: only the latter is convertible, the former is not. This is precisely what the notion of definition as a predication both proper and essential, which expresses an identity both extensional and intensional, comprehending at the same time the objects that are in the extension of the concepts and the meaning of the concepts themselves, allows to technically legislate.[24]

Now, only in the latter case do we have a definition, where both genus *and* differentia are attributed to the subject. In the former case, on the contrary, that which is attributed to the subject is an essential predicate (either the genus *or* the differentia), but not the whole *definiens*.

Thus, in the latter case, the sentence expresses an identity, and is not *strictly* a predication, whereas in the former, despite the fact that the attributed predicates are ἐν τῷ τί ἐστι, they are not the τί ἐστιν itself, and thus the attribution is operated as a predication *stricto sensu* (κατηγορεῖν ἁπλῶς).[25]

---

[24] Cf. *APo.* II 4, 91a15-16; II 6, 92a6-9; II 13, 96a20-b1; but especially: *Top.* I 4, 101b19-23; I 6, 102b27-35; I 8, 103b6-19; VI 1, 139a31-32; VII 5, 154a37-b12; and *passim*.

[25] On the distinction between essential predication and predication of the essence (or definition), the clearest passage by Aristotle is perhaps the following: "For if A is predicable as a mere consequent of B and B of C, A will not on that account be the definable form of C: A will merely be what it was true to say of C. *Even if A is predicated of all B inasmuch as B is precisely a certain type of A* [οὐδ' εἰ ἔστι τὸ Α ὅπερ τι καὶ κατὰ τοῦ Β κατηγορεῖται παντός], *still it will not follow*: being an animal is predicated of being a man (since it is true that in all instances to be human is to be animal, just as it is also

We are now able to establish the sense in which essential predication is and the sense in which it is not strict predication: it is strict predication when that which is attributed is an essential predicate of the subject, but not *the complete essence of the subject*; it is not strict predication when that which is attributed is the very essence of the subject or, in other words, when it is a definition.

One must therefore distinguish between: *(a)* predication of essence; *(b)* essential strict predication ("of the substance"); *(c)* non-essential strict predication ("of accidents"); and *(d)* accidental predication.

In face of these elements, it is now possible to revise the Aristotelian table of predication thus:

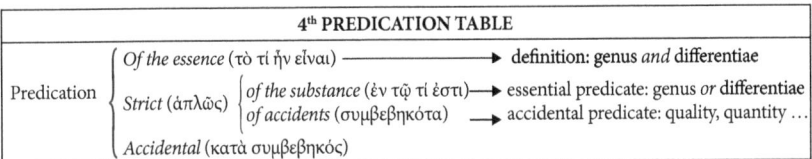

4th PREDICATION TABLE

Predication	*Of the essence* (τὸ τί ἦν εἶναι) ⟶ definition: genus *and* differentiae
	Strict (ἁπλῶς) { *of the substance* (ἐν τῷ τί ἐστι) ⟶ essential predicate: genus *or* differentiae
	of accidents (συμβεβηκότα) ⟶ accidental predicate: quality, quantity …
	Accidental (κατὰ συμβεβηκός)

The following observations may be taken as the key of the table, so to speak:

1) Predication of the essence is the statement of the essence of the subject, i.e., the definition.[26]
2) Strict predication is that in which a "natural" predicate is said of a "natural" subject.
3) When the predicate is part of the subject's essence (in other words, when it is a predicate under the category of substance), the case is one of strict predication of the substance.
4) When the predicate is a simple accidental predicate of subject (alternatively, when it is a predicate under a category other than that of substance), the case is one of predication of the accident.
5) Accidental predication is that in which predication proceeds in a "counternatural" way either because (a), in a sentence, an accident is attributed to another

---

true that every man is an animal), but not identical with being man [ἀλλ' οὐχ οὕτως ὥστε ἓν εἶναι]." (*APo.* II 4, 91b1–7; and cf. also II 13, 96a20-b1)

26 Cf. *Top.* I 5, 101b38. Cf. *APo.* II 3, 90b29–33; *Top.* I 4, 101b17–23; I 8, 103b6–12; V 2, 130b25–28; V 3, 131b37–132a9; VII 3, 153a6–22; VII 5, 154a23–32; VII 5, 155a18–22; *Metaph.* Δ 8, 1017b21–22; Z 4, 1030a2-b13; Z 5, 1031a1–14; Z 13, 1039a19–20; and also *APo.* I 22, 82b37–83a1; II 3, 90b3–4; *Top.* I 6, 102b27–35; I 18, 108a38-b6; V 5, 135a9–12; VI 4, 141a26-b2; VI 4, 141b15–34; *Metaph.* B 3, 998b4–8.

accident ("The musician is white"), or because (b), in a sentence, a substance is attributed to an accident ("that white thing is a log").

As we shall see, the next text will provide us with elements to fine-tune this terminology and to adapt it in accordance with the Aristotelian table of predication.

4. A Preliminary Account

Before moving forwards, though, let us see how these data and those that follow from the previous paragraph enable us to adjust and improve the classification of predication types implicitly addressed in this chapter of the *Posterior Analytics*.

After those paragraphs where he distinguished accidental predication from strict predication, restricted the investigation to the latter and brought back that which can be predicated under the scope of the table of categories, Aristotle advances two steps in this paragraph: on the one hand, he integrates predicates under the category of substance in predication ἁπλῶς; on the other, he reintroduces the notion of accident with a new purpose, viz., not as means to discriminate between types of predication, but to designate one of the predicate classes that, together with those that fall under the category of substance, will exhaust the entirety of what can be strictly predicated.[27]

Taken together, the two newly integrated elements do not add new types of predication to the already established ones. What they do bring is a further characterisation of the types in which strict predication is subdivided: the predication of substance, on the one hand, and the predication of accidents, on the other.

However, the simple fact that Aristotle makes here explicit that predication of accidents is a kind of strict predication is, in itself, significant in another regard.

By doing so, the difference between the two senses in which the word συμβεβηκός may intervene to qualify predication is conclusively rendered clear: in one of those senses, it determines *accidental* predication, which is accidental insofar as it is not predication except by accident (κατὰ συμβεβηκός, *per accidens*); in the other, quite distinct, sense, it delimits predication *of accidents*, which is predication *strictu sensu* (ἁπλῶς), although that which is given through it as predicates of the subject are its accidents (ὡς συμβεβηκότα, *qua accidens*).

In the first case, accidentalness qualifies the very predication: and, via this qualificative, such "predication" stands excluded from the set of strict predication. In

---

27 For which reason, as aptly noted by Ross, "the predication of συμβεβηκότα is of course to be distinguished from the predication κατὰ συμβεβηκός dealt with in the previous paragraph" (Ross 1949: 577).

the second case, accidentalness qualifies but the predicate: therefore, the genuine character of the predication is not affected.

In the first case, accidentalness has a methodological sense and its task is to keep seemingly predicative formulae from the strict domain of predication. In the second case, it bears ontological value and its task is to discriminate a certain type of predicate that has legitimate place in strict predication.

Retrospectively, it is not immaterial that, when distinguishing between accidental predication and strict predication, Aristotle never fails to mention predication of accidents as an instance of strict predication:[28] for that means that, in the distinction between (strict) predication and mere accidental predication, a further distinction, viz., between predication of accidents and accidental predication, is also being established.[29] This is, of course, a particularly important point of the present text.

At the same time, though – as we have just seen in considerable detail –, the paragraph also suggests another relevant aspect: by virtue of its very structure, predication of the substance would be better characterised if we allow it to be distinguished further, between predication of the essence (which is not, strictly speaking, predication, and should therefore be treated separately, viz., as definition) and predication of that which "is in the essence", namely, the genus or the differentiae (which is, from a logical standpoint, strict predication – albeit with unique features – and can thus be considered as a subtype of predication ἁπλῶς, viz., essential predication).

One final observation. Obviously enough, "substance" covers two different meanings in this context: one, the category under which the substance is predicated (i.e., predication of genera or of differentiae); the other, the "natural" subject which, in one of the accidental predication modalities, is shifted to the predicate's logical slot. In neither of these senses, however, is the substance itself a predicate: in the former case, it stands as a category of predicates (the genera and differentiae

---

28 "We can affirm without falsehood that the white (thing) is walking and that that big (thing) is a log; or again, that the log is big and that the man walks. But the affirmation differs in the two cases. When I affirm that the white is a log, I mean that something which happens to be white is a log, not that white is the subject in which log inheres ..." (*APo*. I 22, 83a1-7)

29 The most paradigmatic case is to be found in *APo*. I 19, 81b25-29: "Here is what I mean by 'accidental': when we say, for instance, that that white thing is a man we are not saying the same thing as when we say that the man is white, since the man is not white because he is something else, while the white thing [is a man] because the white is, for man, an accident."

said of subjects); in the latter, it is a substance *strictu sensu*, therefore necessarily a subject that only by accident comes to fill the predicate's logical slot.

One should note at this point that Aristotle does not clarify (a) whether "substance" should be here interpreted as concerning primary substances (in the sense of the *Categories*) only, both primary and secondary substances, or, in general, any subject exhibiting the logical behaviour of a substance,[30] and (b) whether one should take "predicates under the category of substance" to mean those genera and differences said of primary substances only, or these plus those said of secondary substances, or, in general, genera and differences of any subject exhibiting the logical behaviour of a substance.[31]

These questions could have three different answers, depending on the dominion it concerns. In the context of the discussion motivating these developments (viz., the possibility of demonstrations having an infinite number of premises), the appropriate response would be the most restrictive, for the purpose would be that of guaranteeing that the series of subjects stops at individuals (and the series of predicates at categories). In the wider context of *Posterior Analytics*, the convenient answer would be either of intermediate restrictiveness or the broadest possible, in that demonstrations typically deal with universals, for which reason both predicates and subjects should be universal. Generally speaking, nothing militates against choosing the broadest answer; on the contrary, everything points towards it being the favoured one.

5. Predication of Accidents *(a) Per Se* and *(b)* Not *Per Se*

5.1 Text[32]

I assume first that predication implies a single subject and a single attribute [ὑπόκειται δὴ ἓν καθ' ἑνὸς κατηγορεῖσθαι] and secondly that, in the case of non-essential predication, the same things are not predicated of the same things [αὐτὰ δὲ αὑτῶν, ὅσα μὴ τί ἐστι, μὴ κατηγορεῖσθαι]. We assume this because such predicates are all accidents, though some are accidents *per se* [ἀλλὰ τὰ μὲν καθ' αὑτά] and others of a different type [τὰ δὲ καθ' ἕτερον τρόπον]. Yet we maintain that all of them alike are predicated of some subject and that an accident is never a subject, since we do not class anything as accident except when what it says is said due to its being something other than itself [οὐδὲν γὰρ τῶν τοιούτων τίθεμεν εἶναι ὃ οὐχ ἕτερόν τι ὂν λέγεται ὃ λέγεται] ...

---

30 Cf. *Metaph.* Z 1, 1028a36-b2; Z 4, 1030a17-27.
31 Cf. *Top.* I 9, 103b27-39; *Metaph.* Z 1, 1028a36-b2; Z 4, 1030a17-27.
32 *APo.* I 22, 83b17-23.

5.2 Comment

We introduce now the last remaining element that allows us to complete the Aristotelian classification of predication: the distinction between accidents *per se* and "simple" accidents.[33]
We may reformulate the corresponding table thus:

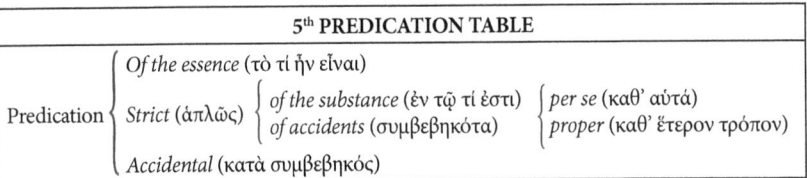

Some final observations in this regard:

1) Predication *strictu sensu* is that which is neither accidental predication nor predication of the essence.
2) It corresponds to three types: *[i]* predication of genus or differences; *[ii]* predication of accidents *per se*; *[iii]* predication of "simple" accidents.
3) "Predication" of the essence is not, in fact, predication, rather an identity formula, in the terms previously mentioned.
4) Accidental "predication" is predication only equivocally: in fact, it occurs only when, by virtue of a grammatical accident, the term that refers to the ontological subject of the attribution slides into the predicate's syntactic slot, which is not naturally its own, so that, in the sentence, there is no actual restitution of any genuine predication.

---

33 Cf. in this regard *Metaph.* Δ 30, 1025a30–34. Other occurrences, both explicit and implicit, of συμβεβηκότα καθ' αὑτά can be found in: *APo.* I 7, 75b1 (and *cf.* I 10, 76b13; I 28, 87a39); *Ph.* II 2, 193b27–28, and III 4, 203b33 (and cf. *De an.* I 1, 402a7; I 1, 402a15; I 1, 402b18; I 1, 402b21; I 1, 402b23–24; I 1,402b26–403a1; I 5, 409b14); *PA* I 3, 643a27–28 (and cf. I 1, 639a18–19; I 1, 639a26; I 1, 641a24–25); *PA* I 5, 645b1–3 (and cf. *HA* I 6, 491a9–11; *MA* 1, 698a1–4); *Metaph.* B 1, 995b20; B 1, 995b25–26; B 2, 997a20; B 2, 997a21–22 (and cf. *Metaph.* B 2, 997a25–34; Γ 1, 1003a21–22; Γ 2, 1005a13–14; E 1, 1025a10–13; E 1, 1026a31–32; K 3, 1061b4–6); M 3, 1078a5–6.

6. In Defence of Aristotle: No Individual Can Be a Predicate

This Aristotelian thesis, which although metaphysical in nature is, as seen throughout the current paper, inseparable from Aristotle's doctrine on predication, was challenged in modern times by several authorities.

Amongst the classic moments, it is worth highlighting those produced during the earlier decades of the 20th century by Frank Plumpton Ramsey,[34] John Cook Wilson,[35] Jan Łukasiewicz[36] and Peter Strawson.[37]

On this particular issue, the work of Cook Wilson is somewhat collateral, since it attempts to present a general doctrine on the nature of the subject and only marginally crosses paths with Aristotle's.[38]

Strawson's essays, in turn, are, to a considerable extent, a recovery of the analysis developed by Ramsey, with whom he would come to part ways later,[39] for which reason it is preferable to resort to the original directly.[40]

We are thus left with Ramsey and Łukasiewicz, to whom we now turn our attention.

The object of Ramsey's essay is to show that "the whole theory of particulars and universals is due to mistaking for a fundamental characteristic of reality what is merely a characteristic of language".[41]

To that effect, he attempts to show that "there is no essential distinction between the subject of a proposition and its predicate", hence "no fundamental classification of objects can be based upon such a distinction".[42]

Ramsey's argument can be schematically presented thus:

---

[34] Ramsey (1925, reedited, with an appendix from 1926, in Braithwaite 1931, from which our quotations are taken).
[35] Cook Wilson (1926).
[36] Łukasiewicz (1957, we translate from the French edition).
[37] Successively in Strawson (1953-1954, 1957a, and 1957b).
[38] In the terms of that doctrine, a subject is that of which a sentence asserts something, which, depending on the context, may or may not coincide with the grammatical subject and, in general, with the nominal component that integrates the sentence.
[39] Cf. Strawson (1959: 177-179 and 237).
[40] For pertinent criticism on Strawson's primitive position, see Sellars (1957) and Baylis (1957).
[41] Braithwaite (1931: 117).
[42] Braithwaite (1931: 116).

1. "Socrates is wise" and "Wisdom is a characteristic of Socrates" express the same proposition.[43]
2. However, that which is subject in one is predicate in the other, and vice-versa.
3. Now, given that any predicative sentence can be analogously converted into an equivalent sentence where subject and predicate switch places, it follows that "there is no essential distinction between the subject of a proposition and its predicate".

The argument would be persuasive if the second premise were true. As it happens, it is not. In fact, the first sentence's predicate is the second sentence's subject, but the first sentence's subject *is not* the predicate of the second. "Socrates" is the subject of the first sentence; but the predicate of the second sentence is not Socrates, it is "a characteristic of Socrates".[44]

Thus, Ramsey only showed something we already knew, at least since Aristotle: that everything that can be a predicate in a sentence can be a subject in another. He did not show, however, what he intended to show: that if it can be a subject in a sentence, then it can be a predicate in another. The Aristotelian irreducibility of the individual as ultimate subject remains thus unscathed.

But there is more. The second premise involves a fatal ambiguity. When we say "that which is subject *in one*, is predicate *in the other*, and vice-versa", the phrases "in one" and "in the other" indicate different things depending on whether they refer, in Ramsey's terms, the sentence or the proposition.[45] The point is that, if one adopts, as does Ramsey, the distinction between sentence and proposition, one assumes that the latter is relatively independent from the former, particularly in view of the fact that it is precisely in order to safeguard the inalterability of the proposition against formal variations that affect the sentence that the distinction itself

---

43 In Strawson, who renovates, although more timidly, Ramsey's argument, the standard example is "All Socrates' virtues were possessed by Plato" (cf. Strawson 1957a: 446–449).
44 Sellars, too, points out, not only against Strawson, but directly against Ramsey, that in "Wisdom is instantiated by Socrates", the predicate is not "Socrates", rather "to be instantiated by Socrates" (cf. Strawson 1957a: 470).
45 The distinction between proposition, the assertive content of a sentence, and sentence, the proposition's material support, can be considered an inextricable topic from essentialism in all its forms, already present in Aristotle, as we had occasion to show in a previous paper (Mesquita 2004: 259–278), and, as we can see, also assumed by Ramsey. Quine made it an irreparably controversial issue; see, especially, Quine (1934; 1958: 21–25; 1960: §§ 40–43; 1968: 139–144; 1970: 1–14; 1974: 36; 1992: 52–53, 77–79, 102; 1995: 77–78).

is put forward. As such, switching the position of the terms within the sentence will not necessarily entail an equivalent permutation in the proposition it expresses, if it is the case, as it is here, that the fact or state of affairs asserted by the proposition remains essentially unchanged when the switch takes place in the sentence. Given that the sentences "Socrates is wise" and "Wisdom is a characteristic of Socrates" express, according to Ramsey, the same proposition, the relation either sentence represents (the fact either sentence asserts, as Ramsey would put it) is the same: the relation of a predicate (wisdom) to a subject (Socrates) – a relation that holds regardless of how the proposition is grammatically transcribed, i.e., whichever term ("Socrates" or "wisdom") happens to be the grammatical subject of the sentence.

Now, there are two important things here. First, when we move from the sentence to the proposition, we realise that the subject *is always the same* – Socrates. That is, at the propositional level (which is to say, at the level of the relation represented by the sentence), the subject is, necessarily, *the ultimate subject*, i.e., the individual. Second, this is precisely why the propositional subject of a sentence "A is B" resists, by nature, being converted into a grammatical predicate; it can but *integrate*, as in the case of Ramsey's example, the grammatical predicate of the converse sentence, which asserts that B is a characteristic of A (or that B pertains to A, or that B is said of A, etc.), i.e., which precisely asserts B as *the predicate* of A.

We can now turn to Łukasiewicz. Commenting on a passage from *Prior Analytics*, he says:[46]

> This passage[47] contains some inaccuracies that it would be best to correct before going any further. It is wrong to say that one thing can be predicated by another; a predicate is part of a proposition, and a proposition consists in a series of uttered or written terms,[48] which possess a certain meaning; therefore, one cannot predicate things: one can predicate the word "Callias" by another word: one cannot predicate Callias himself. The above classification does not concern things – it concerns terms.

The reader who has followed this paper thus far will realise that this excerpt is built upon a fundamental misunderstanding concerning two levels that Aristotle keeps quite apart: the ontological level, where he speaks of predicates as something that pertains *to* things; and the logical level, where he speaks of predicates as something that is said *of* things (through the terms by which things are referred to). In the latter case, the predicate is in fact a term and it is in fact part of a sentence; in the

---

46 Łukasiewicz (1957: 26).
47 *APr.* I 27, 43a25–43.
48 "Proposition", for Łukasiewicz, has of course the same meaning as "sentence".

former, however, the predicate is *an entity* and, as such, it is utterly extra-logical and extra-linguistic. The entire Aristotelian system of categories (the ways according to which *something is said to be*) can only make sense assuming this presupposition. More than that: the two levels are connected – and they are connected by *the primacy of the former over the latter*. In fact, it is only because something pertains to something else *qua* an ontological predicate that it can be said of that thing as a logical predicate. It is only insofar as something is *a predicate of something* that it can become *that which is predicated* of that thing in a sentence (in the latter case, subject and predicate are not, of course, the entities themselves, rather the terms that refer to those entities).[49]

In this sense, Aristotle never states or implies that "one thing can be predicated by another": what he does say is that "something" (i.e., an entity, in general terms) can be *a predicate of* another. He certainly states and implies, however, that "things can be predicated": for, although the subject of a predicative sentence is not the thing itself, it is the thing itself that, through its name, is predicated by the predicate of the predicative sentence.[50]

On our subject, Łukasiewicz adds:[51]

> Likewise, it is wrong to say that individual or singular terms, e.g., "Callias", cannot be truly predicated of anything. Aristotle himself is the first to line up examples of true propositions with singular predicates: "This white object is Socrates"; or "That who approaches is Callias". These propositions are true, he says, "by accident",[52] but there are other examples

---

49 On the problem of accidental predication, Lear sees, correctly, the juxtaposition of two levels – logical and ontological – in all predication. Cf. Lear (1980: 31): "A phrase like 'the white thing is a log' is a degenerate form of predication, for it fails to reveal the metaphysical structure of subject and predicate. It is not that the white thing is the underlying subject which happens to be a log. Rather the log is the underlying subject which happens to be white (*An. Pst.* 83a1–14). Only predications which reveal metaphysical structure are strict and it is with these that proof is concerned." And a few lines ahead: "Aristotle distinguishes predicating from saying truly (*An. Pst.* 83a38). Predication is not merely a linguistic act. Though one can say both 'the white thing is a log' and 'the log is white', only the latter is a genuine predication."
50 In fact, even from the point of view of modern elementary logic, what does it mean to say that an object satisfies a predicate, other than that object *has* the property signified by the predicate and, consequently, that the predicate is predicated *of it*? Here is a point concerning which there has been a rather unjustified desire to draw distinctions between the assumptions of modern logic and those of Aristotelian logic.
51 *Op.* and *loc. cit.*
52 Łukasiewicz has here in mind those cases of accidental predication (κατὰ συμβεβηκός) to which we have made abundant reference in the text.

of the same kind which truth is not purely accidental, such as "Socrates is Socrates" or "Sophroniscus was Socrates' father".

Again, not only this excerpt depends on the misconception just detected – for the notion of predication lies, as we saw, upon an *ontological* distinction that Łukasiewicz misses –, it also is underpinned by four examples *none of which is an instance of predication*, due to the exact reasons that our preceding analysis has made clear.

This is evident in the case of the first two examples, taken from the *Prior Analytics* chapter under debate, for they correspond precisely to the examples given for accidental predication.

In "that white object is Socrates", as well as in "that who approaches is Callias", "Socrates" and "Callias" are not predicates of "that white object" and "that who approaches", respectively; rather, they are their corresponding *names*. The sentence is, consequently, an identity formula – not a predication.[53]

As to the third example, "Socrates is Socrates", it is difficult to see how the second "Socrates" could be a predicate of the first. Why is the second Socrates said of the first and not the other way around? Could the author be suggesting that it is the order of the sentence that determines the predicate of the attribution? But then, if one says "Socrates was a great philosopher", "Socrates" is the subject, yet if one says "A great philosopher was Socrates", "Socrates" becomes the predicate? It is not, evidently, the order of the sentence that determines the predicate of the attribution. And it is also clear that the first and the second "Socrates" in the sentence "Socrates is Socrates" are undistinguishable as subject and predicate, which means, in other words, that the sentence does not express a predication.

The fourth example is the most surprising. How does Łukasiewicz interpret the clause "Socrates' father" in the sentence "Sophroniscus was Socrates' father"? As a singular term? If it is a singular term, then it is either a name or a definite description: and, in both cases, not a predicate. Conversely, if it is a predicate (the predicate "*x* is Socrates' father"), then it is not a singular term. In the first case, we have no predication; in the second, the predicate is not an "individual or singular term", in Łukasiewicz's terms. In both cases, the example does not demonstrate what Łukasiewicz intended it to demonstrate.

---

53 Cf., along the same line, Smith (1995: 33): "But when we force 'Socrates' into predicate position, what we have no longer seems to be predication, but instead a kind of identification: 'That man is Socrates' amounts to 'That man and Socrates are the same.'"

The crucial point here is that Łukasiewicz's whole reasoning seems to miss – fundamentally – the Aristotelian doctrine in question.

What Aristotle implies in the notion of accidental predication is the *ontological* thesis according to which an individual cannot be a predicate of another. It is not a question of terms or linguistic predicates: it is a question of real things and its actual predicates, in Aristotle's primary ontological sense.

Now, thus understood, the doctrine is clear as to all the examples given: the individual Socrates is not a predicate of that white blotch that I see over there; rather, white is a predicate of the individual Socrates. And it is only by virtue of a linguistic accident that we can *incorrectly* express this truth, saying "that white thing is Socrates".

From a logical point of view, the ontological notion of accidental predication means, thus, the following: the "correct" subject of the sentence, that is, the term that refers to the subject of the attribution, shifts, by linguistic accident, into the predicate's slot. And this entails, in the terms above mentioned, a more fundamental idea: the *sentence* "that white thing is Socrates" expresses the *proposition* "Socrates is white". (Naturally, the same is valid in the sentence "that who approaches is Callias" and other similar sentences.)

In the remaining examples, which are not instances of accidental predication, the ontological thesis according to which no individual can be a predicate is also not undermined. In fact, all these (and the same would hold for the previous examples, which can be likewise interpreted) are cases of identity, not predication, and thus none exhibits individuals *qua* predicates.

Alternatively, one would have to admit that, in sentences like "Sophroniscus was Socrates' father", the predicate ("Socrates' father") is not a singular term, but, rather surprisingly, a general term (in which case the sentence should seemingly be read as "Sophroniscus was *a* Socrates' father") and therefore the Aristotelian thesis is, once again, not affected.

7. Final Discussion and Conclusions

The Aristotelian notion of accidental predication appeals, first of all, to a question of purely ontological character – one which is crucial to acknowledge as such, in order to avoid mistaking it with the related logical question.

The ontological issue is thus: there are certain entities that, due to their very nature, cannot be predicates of anything.

Łukasiewicz saw fit to correct Aristotle, recalling that only terms – not entities – can be predicates. But this only manifests the confusion as to the two levels. In Aristotle's thought, a predicate is one thing and that which is predicated (viz., in the predicative sentence) is a whole other thing. *Predicate* is an ontological notion, and it has to do with that which belongs to a given thing (thus called, in equally ontological manner, its subject). *That which is predicated* is a logical notion, strictly dependent from the ontological notion, and it has to do with what can *be said* of the subject (in a predicative sentence). In a predicative sentence, that which is predicated is, of course, a term. But this term can only be predicated of another because what it designates *is a predicate* of that which is designated by the other.

Now, there are things that cannot be (ontological) predicates of anything – such is the case of individuals. Accordingly, if the name of one of those things (i.e., a singular term) comes to be a part of a predicative sentence, filling the predicate's slot, there we have, literally, *an accident*.

But do such cases really exist? Surely not in Aristotle's examples ("The musician is white", "that white thing is a log"): in his examples, what fills the predicate's slot is never a singular term.

However, for the sake of the argument, one could re-read the examples in Łukasiewicz's vein ("this white object is Socrates", "that who approaches is Callias"), where what fills the predicate's slot is, surely enough, a singular term.

Before this situation, we are forced to reiterate: only accidentally can a name (a singular term) appear as a predicate in a predicative sentence, since the individual to which that name corresponds *is not the actual predicate* of the thing referred to by the term appearing as subject in the sentence, as it is patently shown in Aristotle's examples – and in Łukasiewicz's, for that matter – and well explainable in light of the Aristotelian doctrine.

Now, that accident by which the name of something that cannot be a (ontological) predicate appears as (logical) predicate is, quite clearly, an accident of linguistic nature.

This is what Aristotle is saying when he states that the sentence "this white thing is a log" is an inversion of the true predication "the log is white". And this is also the notion present in the neoplatonic distinction between "natural" and "counternatural" predication.

It is in this case, and in this case only, that the second level – the logical plan – intervenes. One could argue that, from a logical standpoint, the distinction between natural and counternatural predication (an ontological distinction) is meaningless and, thus, that it is meaningless to say that the "real" subject of the proposition expressed by the sentence "this white thing is a log" is that which is represented, in the sentence, by the predicate. From a strictly logical point of view, "log" is said of "this white thing" and is, therefore, the sentence's predicate. As to those ontological claims according to which, in the order of reality, it is white which is the predicate of log, and not the other way around, logic may very well respect them, but it is under no obligation to take them into consideration.

However, if this were the case, in order to accommodate *in extremis* Aristotle's thesis, it would be necessary to concede that, although the objects it designates could never be (ontological) predicates, singular terms themselves could, indeed, be (logical) predicates.

As it happens, this is not so.

First of all, an Aristotelian could always argue that logic does not deal with sentences; it deals with propositions, where a singular term is *never* a predicate (except, of course, in identity formulae). Proposition, the last stronghold of essentialism, is also the last stronghold of metaphysics; and in the intensional dwelling of propositions, the metaphysical order of reality can always be preserved.

Secondly, Aristotle would still have good reasons to maintain that, even from a logical standpoint, singular terms *cannot* be predicates. Why? Because, in every situation where accidental predication occurs, either the predicate is not an actual singular term, or *the case is not even one of predication*.

In "this white thing is a log" (a typical example in Aristotle), "a log" most certainly is not a singular term – it is a general one. What the sentence means to express is that it happens to this white thing (singular) to belong to the class (universal) of logs. (To be perfectly fair to Aristotle, one should have put it vice-versa, but, for our purposes, the warning will suffice.)

No doubt this sentence, and every sentence equivalent to it, can always be read placing a singular term in the predicate's position. In such case, it would read something like "this white thing is this log" – graceless equivalent of "this white thing is Socrates" or "that there is Callias". But then, "this log", or "Socrates", or "Callias", is not attributed to "this white thing" as predicate, rather as another name for "this white thing", in which case the sentence is not asserting predication, but indeed identity, between the two terms.

In this case, Aristotle's accidental predication would be assimilated into the second class of Łukasiewicz's examples. In other words, Aristotle's accidental predication would be systematically reinterpreted as an ill-formed (or "accidental") identity formula.

It is, to be sure, a somewhat far-fetched hypothesis.

Yet, it holds a certain appeal, for it would allow us to reunite anew the two fringes, "lower" and "upper", of Aristotle's predication.

References

Barnes, J. (tr.) (1994). *Aristotle. Posterior Analytics* (2nd ed.). Oxford: Clarendon Press.

Baylis, C. A. (1957). "Logical Subjects and Physical Objects. Comments". *Philosophy and Phenomenological Research* 17: 483–487.

Bonitz, H. (1961). *Index Aristotelicus, Aristotelis Opera. Editio altera*, ed. I. Bekker, revis. O. Gigon, V, Berlin: Königlichen Preussischen Akademie der Wissenschaften.

Braithwaite, R. B. (1931). *The Foundations of Mathematics and other Logical Essays*. London: Kegan Paul, Trench, Trubner & Co.

Cook Wilson, J. (1926). *Statement and Inference*, I-II. Oxford: Clarendon Press.

Lear, J. (1980). *Aristotle and Logical Theory*. Cambridge: Cambridge University Press.

Łukasiewicz, J. (1957). *Aristotle's Syllogistic from the Standpoint of Modern Formal Logic*, (2nd ed.). Oxford: Oxford University Press (French edition: tr. F. Caujolle-Zaslawsky, Paris, Librairie Armand Collin, 1972).

Mesquita, A. P. (2004). *Aspectos Disputados da Filosofia Aristotélica*. Lisboa: INCM.

Mure, G. R. G. (tr.) (1908). *The Works of Aristotle translated into English*, vol. I. Oxford: Clarendon Press.

Pellegrin, P. (tr.) (2005). *Aristotle. Seconds Analytiques*. Paris : Garnier-Flammarion.

Quine, W. V. O. (1934). "Ontological Remarks on the Propositional Calculus". *Mind* 43: 472–476 (reedited in: *The Ways of Paradox and Other Essays*, Cambridge, MA and London: Harvard University Press, 1997, 7th ed., 265–271).

Quine, W. V. O. (1958). "Speaking of Objects". *Proceedings and Addresses of the American Philosophical Association* 31: 5–22 (reedited in: *Ontological Relativity and Other Essays*, New York: Columbia University Press, 1969, 1–25).

Quine, W. V. O. (1960). *Word and Object*. Cambridge, MA: MIT Press.

Quine, W. V. O. (1968). "Propositional Objects". *Crítica* 2: 3–22 (reedited in: *Ontological Relativity and Other Essays*, New York: Columbia University Press, 1969, 139–160).

Quine, W. V. O. (1970). *Philosophy of Logic*. Englewood Cliffs N. J: Prentice Hall.

Quine, W. V. O. (1974). *The Roots of Reference*. LaSalle, Ill: Open Court.

Quine, W. V. O. (1992). *Pursuit of Truth* (2nd ed.). Cambridge, MA and London: Harvard University Press.

Quine, W. V. O. (1995). *From Stimulus to Science*. Cambridge, MA and London: Harvard University Press.

Ramsey, F. P. (1925). "Universals" *Mind* 34 (reedited in: *The Foundations of Mathematics and other Logical Essays*, ed. R. B. Braithwaite, pref. G. E. Moore, London: Kegan Paul, Trench, Trubner & Co., 1931, 112–134, 135–137).

Ross, W. D. (ed.), (1949). *Aristotle's Prior and Posterior Analytics. A Revised Text with Introduction and Commentary*. Oxford: Clarendon Press.

Sellars, W. (1957). "Logical Subjects and Physical Objects". *Philosophy and Phenomenological Research* 17: 458–472.

Smith, R. (1995). "Logic". In: J. Barnes (ed.), *The Cambridge Companion to Aristotle*, Cambridge: Cambridge University Press, 27–65.

Smith, R. (tr.) (1994). *Aristotle. Topics Books I and VIII*. Oxford: Clarendon Press.

Strawson, P. F. (1959). *Individuals. An Essay in Descriptive Metaphysics*. London: Methuen.

Strawson, P. F. (1957a). "Logical Subjects and Physical Objects". *Philosophy and Phenomenological Research* 17: 441–457.

Strawson, P. F. (1957b). "Logical Subjects and Physical Objects. A reply to Mr. Sellars", *Philosophy and Phenomenological Research* 17: 473–477.

Strawson, P. F. (1953-1954). "On Particular and General". *Proceedings of the Aristotelian Society* 54: 233–261 (reedited in: *The Problem of Universals*, ed. C. Landesman, New York: Basic Books, 1971, 131–149).

Tredennick, H. (ed.) (1989). *Aristotle* II. London: Loeb Classical Library.

Tricot, J. (trad.) (1965). *Aristotle. Organon. IV: Les secondes analytiques*. Paris : Vrin.

Waitz, T. (ed.) (1844–1846). *Aristotelis Organon graece*, I–II. Lipsiae, sumptibus Hahnii.

Wallies, M. (1960). *Commentaria in Aristotelem graeca, edita consilio et auctoritate Academiae Litterarum Regiae Borussicae*, XIII.3. Berlin: G. Reimer.

Nino B. Cocchiarella
Indiana University

On Predication, A Conceptualist View

Abstract: Predication, as the nexus between a subject and a predicate expression, is the basis of the unity of a speech act, including speech acts in the plural and speech acts that involve mass nouns. A speech act, of course, is an overt expression of a mental act, e.g., a judgment; and therefore the unity of a speech act such as an assertion is really the unity of the judgment that underlies that act. Such a mental act, and therefore the speech act as well, has a unity based on how the referential and predicable roles of the subject and predicate expressions combine and function together respectively. What we propose here is to explain this unity of predication in terms of a conceptualist theory of logical forms that we claim underlies at least some important aspects of thought and natural language. Our conceptualist logic also contains an account of the medieval identity (two-name) theory of the copula, as well as an account of plural and mass noun reference and predication, the truth conditions of which are based on a logic of plurals and mass nouns.

Keywords: predication, reference, pluralities, mass nouns, Conceptualism, cognitive capacities, copula, unsaturatedness, Medieval Logic,

0. Introduction

A speech act such as an assertion is not just a jumble of words, nor is the judgment that underlies it a stream of mental flotsam. An assertion and the judgment underlying it have a unity, something that accounts for the coherence of the judgment and the assertion it underlies. What is the basis of that unity or coherence? How can we account for it? It cannot be the rules of grammar that tells us how the words and phrases are to be correctly put together so as to express a judgment or a thought. Most people do not know the rules of grammar even after they have become competent in speaking and expressing their thoughts. So just what does hold the parts of an assertion or of a thought together?

As a speech act, an assertion is an overt expression of a mental act, i.e., of a judgment, and therefore the unity or coherence of such a speech act is really the unity or coherence of the mental act that underlies it. This unity, we claim, is based on how the referential and predicable concepts of the subject and predicate expressions of such an act combine and function together. Such a combination and mutual functioning of referential and predicable concepts is what we mean here by predication. What we propose is to explain the unity of predication in terms of a conceptualist theory of logical forms that represent referential and predicable concepts and that we believe explains at least some important aspects of thought and natural language. Our conceptualist logic also contains an account of plural and mass noun predication, i.e., of plural and mass noun reference and predication, the truth conditions of which are based on a logic of plurals and mass nouns that we have developed elsewhere.[1]

Concepts in this conceptualist theory are the intersubjective cognitive capacities humans acquire in learning a language and that subsequently come to underlie our use of language. Predicable concepts, in particular, are the cognitive capacities underlying our rule-following use of predicate expressions, and referential concepts are the cognitive capacities underlying our rule-following use of referential expressions. As cognitive capacities, concepts are the basis of a kind of pragmatic knowledge, or more specifically a knowing how to do things with referential and predicable expressions in various contexts of use of language and thought. Having such concepts is not a form of propositional knowledge, i.e., knowledge that certain rules of language are correct, even though having such concepts underlies the rule-following behavior those rules might describe. As intersubjectively realizable cognitive capacities that might be exercised at the same time by different persons and at different times by the same person, referential and predicable concepts are in an appropriate sense objective cognitive universals.

According to Kant, what unifies a mental act is a "synthetic unity of apperception of the thinking subject".[2] Such an apperception, Kant claimed, not only unifies each judgment, but it also determines the categorial structure of the different possible judgments that can be made. In fact, according to Kant, what categories there are and how they fit together is determined by a "transcendental deduction", and the categories so deduced form the basis of a so-called *transcendental logic*. But just how the synthetic unity of apperception unifies a judgment is not clear. Apparently, it is assumed that the unity is somehow determined by the mind of the thinker. The question then is just how does the mind do this? Is there a way we can logically represent this unity? We believe that the theory that we propose here for our conceptualist logic might possibly be taken as providing an answer to this last question, but in no sense do we claim that our logic is based on either a transcendental deduction or a synthetic unity of apperception.

Some philosophers claim that predication can be explained only in terms of a relation or "tie" of exemplification. Exemplification is said to be what holds

---

1 See Cocchiarella (2002) and (2009).
2 Kant (1965, §16), "The Original Synthetic Unity of Apperception".

together (and hence unifies) what the subject of a sentence denotes with what the predicate stands for, the result being a state of affairs or a proposition. For the sentence 'Socrates is wise', for example, exemplification is what relates or ties Socrates with wisdom. Wisdom of course is an abstract entity that does not exist in the world the way Socrates does (or did). That is, wisdom exists – or rather has *being* if *existence* applies only to things in the physical world – in some abstract platonic realm. Exemplification, accordingly, is assumed to be a relation or "tie" between something concrete or physical and something abstract. Aside from the question of how there could be a such a relation or "tie" between objects of different realms or modes of being, this sort of answer does not explain how the referential and predicable concepts that are exercised in a judgment are unified in a speech or mental act. After all, a mental act, such as a judgment, does not have either a concrete object such as Socrates or an abstract object such as wisdom as a constituent. A relation of exemplification, in other words, is not what explains the unity of a speech or mental act.

Some other philosophers, including especially Gottlob Frege, would object to a relation of exemplification because it replaces the predicate phrase, e.g., 'is wise', by its nominalized form, 'wisdom', an objection with which we agree. The predicate phrase in 'Socrates is wise' does not function as a name, in other words, and that is because what the predicate stands for is not an object (or thing) that can be named. According to Frege, what a predicate stands for is an abstract unsaturated function from objects to truth values, which Frege also called a concept, but not in the sense of a cognitive capacity. Instead of exemplification, what the sentence 'Socrates is wise' says is that Socrates falls under the abstract concept that the predicate stands for, i.e., the abstract function that 'is wise' stands for assigns truth to Socrates. This answer is also unsatisfactory, because as an abstract entity a Fregean concept is not a mind dependent entity, and therefore it has nothing to do with the unity of a mental act.

Finally, whether predication is explained in terms of exemplification or an abstract Fregean function from objects to truth values, many philosophers view the judgment that 'Socrates is wise' as expressing a belief, which is then assumed to be a relation between a person's mind and a proposition (or state of affairs) having Socrates and the property of being wise as constituents. Be that as it may, such a view still does not account for the unity of the judgment as a speech or mental act, and therefore we will not be concerned with such a view here, whether positively or negatively.

These kinds of answers, and others like them, are semantical accounts of the truth conditions of a speech act, and in particular a speech act of rather simple form where the subject is either a proper name or a definite description. None really explains (or purports to explain) what accounts for the unity of a speech or mental act. What these sorts of theories purport to explain are the truth conditions of simple subject-predicate sentences, from which, supposedly, the truth conditions of other kinds of sentences are then determined as in a model-theoretic semantics. We do not dispute the importance of such an account of truth conditions, but whatever the significance of this approach it does not explain the problem of the unity of a speech or mental act. We do assume, as far as truth conditions are concerned, that the logical forms of our conceptualist theory provide a perspicuous guide to what logically follows from what, and in that regard the theory provides all of the semantics that we need for our present purpose. In other words, we assume that the logical forms of our underlying logic carry their semantics on their sleeves.

1. General Reference and the Medieval *Suppositio* Doctrine

Subject expressions of natural language, and of English in particular, consist not only of proper names or definite descriptions, but also of quantifier phrases such as 'Every republican who voted for the bill' and 'Some democrat who did not vote for the bill'. As subject expressions of speech acts, these quantifier phrases, are used to express different forms of general reference, and the unity of a speech act based on the use of such an expression must be accounted for no less so than those having a proper name or definite description as a referring subject expression. Our theory, in other words, must account for general reference as well as singular reference. Here again one must not confuse accounting for the truth conditions (as in standard model theory) of these general statements or judgments in terms of their instances with explaining their unity as speech acts or judgments. That in effect amounts to changing the subject.

Historically these kinds of speech and mental acts were taken seriously by medieval logicians and linguists, especially in the 14th c. They were explained primarily by the medieval *suppositio* (supposition) theory of William of Ockham, John Buridan, Walter Burley, Gregory of Rimini, and a number of others. Today this theory is generally called *terminist logic* because of its focus on the categorematic expressions of Latin that were called *terms*. As in our conceptualist theory, reference in terminist logic was a pragmatic notion, i.e., it was intended to apply to our speech and mental acts in different contexts of use.

One important part of *suppositio* theory is the doctrine of the modes of supposition. This doctrine, despite the reference to "modes" of *supposition*, is not a theory about different "ways of referring". Rather, the "modes" are just different types and subtypes of *personal supposition*, which is reference to things. We have explained elsewhere how these different subtypes of personal supposition, and Ockham's *suppositio* theory in general, can be reconstructed in terms of our conceptualist theory of reference, and the details of that analysis will not be repeated here.[3] Instead, we briefly consider the kind of language that was implicit in terminist logic, and then in a later section discuss the terminist identity (or two-name) theory of the copula, which will be useful in our account of predication.

Implicit in terminist logic was the view that there is a language of thought that today is sometimes called *Mental*, a language that was made up of both categorematic and syncategorematic concepts.[4] This mental language was assumed to be a "natural language" common to all humans and somehow established by nature. What makes Mental natural is that its categorematic concepts (mental terms) get their signification (reference) by nature and not by convention. The assumption was that there is a "natural likeness" between concepts and the things they signify (refer to), a likeness that was caused by the things signified.

Implicit in this theory was the idea that concepts are like images that resemble things, a view that is quite different from the kind of cognitive capacities that we take them to be in our form of conceptualism. For a variety of reasons that we will not go into here, this "imagist" theory is an inadequate account of concepts. One reason is that many of the concepts that are cognitive capacities in our form of conceptualism, such as the concepts we have in mathematics, and the various concepts we have about the microphysical world, can in no sense be said to have a likeness to things in nature.[5] Otherwise than this sort of difference in the nature of concepts much of what the medieval philosophers and linguists have to say about Mental can be explained in terms of the logical framework of our form of conceptualism.

Mental is a tensed and modal language containing among its syncategorematic concepts certain operators that correspond to the tenses and modal modifications of verbs, or what the medieval logicians called ampliation. Our conceptualist logic includes these features and represents them as formula operators, but for convenience we will not deal with the details of tense and modal logic in this paper – except briefly later in order to deal with a specific problem regarding predication based on relations. Also, because of ampliation mental terms were taken to signify (refer) in a wider sense than just to (presently) existing things. In

---

3   See Cocchiarella (2001). See also Spade (2007).
4   See, e.g., Geach (1980b), Normore (1990), and Trentman (1970). See also Scott (1966b).
5   For more on this difference between *Mental* and our form of conceptualism see Cocchiarella (2001, §2).

other words, Mental was ontologically committed not just to present but also to past and future objects, and even possible objects. The possible objects that were signified (in the wider sense), however, seemed to be only those that are possible in nature.[6] In our form of conceptualism we also allow for the kind of possibilia we believe to be implicit in natural language, but we distinguish the many concepts that entail existence such as *dog*, *house*, *man*, etc., and even *dragon*, from those that do not. In other words, something cannot be a dog, house, man, or a dragon unless it exists. But then we also distinguish fictional and mythological contexts from ordinary contexts of use, so that although no dragons exist in reality, some dragons exist in fiction.[7]

According to Geach, the "grammar of Mental turns out to be remarkably like Latin grammar", and indeed Ockham did seem to carry over some of the features of Latin into Mental.[8] But according to J. Trentman, "Ockham's real criterion (…) for admitting grammatical distinctions into Mental amounted to asking whether the distinctions in question would be necessary in an ideal language – ideal for a complete, true description of the world".[9] Similarly, according to many medieval specialists, the proper comparison is not of Mental with Latin but of Mental with the kind of "ideal languages" that logicians and philosophers have constructed in the twentieth century, i.e., with the kind of logical system such as our conceptualist theory of logical form.

What is needed for an adequate interpretation of terminist logic is a representation of Mental as a logistic system, and in particular one based on a view of logic as language as opposed to logic as calculus. We take the logic of our conceptualist theory of predication to be just that sort of system. We turn now to the first part of our version of such a logic.

2. General Reference in the Logic of Names

The initial form of our theory of reference is called a logic of names, where by names we mean not just proper names but also common nouns, whether simple or complex, and even verbal nouns complex or simple. Verbal nouns are used to refer to events, as when we speak of a kissing, or a jumping, etc. For convenience

---

6 See Normore (1985: 191). We note that a natural possibility and necessity seems to be what Burley had in mind.
7 See Cocchiarella (2013) for more on the distinction between existence-entailing concepts and concepts that do not entail existence.
8 See Geach (1980b).
9 See Trentman (1970: 589).

we will not deal with verbal nouns in this paper. In regard to names otherwise than verbal nouns, we note that medieval terminist logicians held, as did Leśniewski in his logic of names (which he also called ontology), that the category of names includes common names, i.e., common nouns, as well as proper names. This is different from most views today where common names are taken as predicates and proper names, along with definite descriptions, are taken as singular terms, which is an entirely different category from predicates. Peter Geach, incidentally, changed his mind on this matter and came to agree with Leśniewski's and our view of names. Geach, in other words, "came to accept the view of Leśniewski and other Polish logicians that there is no distinct category of proper names (....) in syntax there is only the category of names".[10]

We take it that a basic form of judgment is expressed by an assertion that consists of a noun phrase and a verb phrase, and that the noun phrase has a referential role regardless whether or not it is a proper name, a definite description, or a quantifier phrase. A definite description is viewed in our theory as a quantifier phrase on a par with a universal or existential quantifier phrase.[11]

But because our first-order logic is free of existential presuppositions, a definite description may fail to refer to anything, as might the use of a proper name as well.

A quantifier phrase in our logic of names is made up of two parts, the first being a determiner such as 'every', 'some', the indefinite article 'a', and the definite article 'the' – and others as well, such as 'most', 'few', etc., which we will not deal with here. The second part of a quantifier phrase is a name in our present sense, which may be a proper name or a common name, i.e., a common noun, which could be a mass noun, a count noun, or a gerund in its role as a verbal noun (and which, as noted, we will not deal with here). A count noun can be simple, such as, e.g., 'politician', or complex, such as 'politician who is conservative', where the complexity is the result of affixing a qualifying relative clause, such as 'who (or that, or which) is conservative', to the head noun. Similarly, a mass noun can be simple, such as, e.g., 'water', or complex as with 'water that is polluted'.

Our basic logic, which we call a logic of names, is an extension of free first-order predicate logic with identity, i.e., first-order predicate logic free of existential presuppositions. The extension consists of adding a new syntactic category, which we call the category of names (in our present sense). The name variables of this category may be bound by either the universal or existential quantifiers. Name variables and constants are the simple names in this category. Complex names

---

10 Geach (1980b: 15).
11 Evans held a similar view of definite descriptions in (1982: 57).

are formed by means of a relative-clause operator. Leśniewski's logic of names is different from ours, but, as we have shown in Cocchiarella (2001), Leśniewski's logic of names is reducible to our logic of names, i.e., Leśniewski's logic can be translated into our logic so that every theorem of his system is a theorem of ours.[12]

Our logic of names contains absolute (unrestricted) as well as restricted, relative quantifier phrases, i.e., relative quantifier phrases such as $(\forall xA)$ and $(\exists xA)$, read as 'every A' and 'some A', respectively, where A is a name, common or proper, and complex or simple.[13] We will use the standard quantifier forms $(\forall x)$ and $(\exists y)$ for the absolute quantifier phrases. We use x, y, z, etc., with or without numerical subscripts, as first-order variables and A, B, C, with or without numerical subscripts, as name variables. Complex names are formed by adjoining defining or restricting relative clauses to names. For our relative-clause operator we use '/', as in $A/\varphi x$, to represent the adjunction of a formula φx to the name A (which may itself be complex). We read $A/\varphi x$ as 'A that is φx'. Thus, e.g., the quantifier phrase representing reference to *a brown dog* can be symbolized as $(\exists xDog/Brown(x))$. We assume that an attributive adjective, such as 'brown' in 'brown dog', is equivalent to its occurrence as a predicate in a relative clause, as in 'dog that is brown'. Attributive adjectives such as 'alleged' in 'alleged thief' are really operators, as in 'person who is alleged to be a thief', where 'alleged' would be symbolized as the operator 'it is alleged that'.[14] A relative clause might itself be complex, of course, which is then represented by an iteration of the operator /. The complex quantifier phrase 'a brown dog that is vicious', for example, can then be represented as $(\exists xDog/Brown(x)/Vicious(x))$.

A proper name, such as 'Gina', for example, can occur in our logic as part of a quantifier phrase, as in $(\exists xGina)F(x)$, which indicates that the name 'Gina' is being used with existential presupposition. In our logic of plurals (and mass nouns), which we describe later as an extension of our logic of names, all names, proper or common, can be transformed into "terms", i.e., arguments of predicates, and one result of such a transformation is that the formula $(\exists xGina)F(x)$ turns out to be equivalent

---

12 See Cocchiarella (2001) for a proof of this reduction.
13 Absolute quantifiers will range over not just every object or thing, but every plurality as well, or what later we will call classes as many. As we note later, this will include what mass nouns denote as well.
14 Some attributive adjectives such as 'big' and 'small', as in 'big mouse' and 'small elephant' have an analysis more involved than as simple predicate adjectives; but we will not deal with those adjectives here.

to the more standard free-logic expression.[15] Thus whereas $(\exists xGina)F(x)$ represents the referential role of the name 'Gina' in a speech act, the formula $(\exists x)[x = Gina \wedge F(x)]$ represents the truth conditions and logical implications of that speech act.

As already noted, definite descriptions are also quantifier phrases in this logic no less so than indefinite descriptions. As quantifier phrases, both definite and indefinite descriptions can be used as referential expressions, though of course they differ logically (and therefore semantically) in their referential roles. Definite descriptions, as indicated, can also be used with or without existential presupposition. We will use $\exists_1$ for the definite description operator when it is used with a presupposition and $\forall_1$ when without. Our analysis of a use of $\exists_1$ agrees in essentials with Bertrand Russell's contextual analysis of definite descriptions. We symbolize 'The A is F' as $(\exists_1 xA)F(x)$ when 'The A' is used with existential presupposition, and as $(\forall_1 xA)F(x)$ when used without.[16] Thus, e.g., an assertion of 'The black dog is vicious', where the definite description is being used with existential presupposition is symbolized as:

$(\exists_1 xDog/Black(x))Vicious(x)$,

which represents the form of the speech act, whereas

$(\exists xDog/Black(x))[(\forall yDog/Black(y))(y = x) \wedge Vicious(x)]$

gives a more perspicuous representation of its truth conditions. Applying meaning postulates about how relative clauses and relative quantifiers are to be expanded, this last can be further expanded as follows:[17]

$(\exists x)[(\exists yDog)(x = y) \wedge Black(x) \wedge (\forall z)((\exists yDog)(z = y) \wedge Black(z) \rightarrow z = x) \wedge Vicious(x)]$.

---

15 This use of "term" should be distinguished from what the medieval logicians meant by a term.
16 The truth conditions of these formulas are given more perspicuously as,

$(\exists_1 xA)F(x) \leftrightarrow (\exists xA)[(\forall yA)(y = x) \wedge F(x)]$,

for situations where the definite description is used with existential presupposition, and as

$(\forall_1 xA)F(x) \leftrightarrow (\forall xA)[(\forall yA)(y = x) \rightarrow F(x)]$

where used without.
17 The meaning postulates in question here are as follows:

$(\forall xA/F(x))G(x) \leftrightarrow (\forall x)[(\exists yA)(x = y) \wedge F(x) \rightarrow G(x)]$

and

$(\exists xA/F(x))G(x) \leftrightarrow (\exists x)[(\exists yA)(x = y) \wedge F(x) \wedge G(x)]$.

This formula gives even more details about the truth conditions in question. But these are matters about the background logic and truth conditions of our speech and mental acts, as opposed to the logical forms of our speech or mental acts themselves as described in terms of the referential and predicable concepts being exercised. In other words, the first of the above formulas represents the logical form of the speech act in question, namely the assertion that the black dog is vicious, whereas the last provides a more perspicuous representation of its logical form with respect to its deductive implications, and therefore its truth conditions as well. This distinction between the logical forms that represent the mental or speech-act level and the logical forms that represent the deductive truth-conditional level is a fundamental feature of our conceptual logic, and it is important that it be kept in mind in understanding how this theory works.

Finally, we should note that *Dog* is not a predicate in our logic but a common name, and $Dog(x)$ is neither well-formed nor the proper way to symbolize the statement that x is a dog. Instead we use $(\exists y Dog)(x = y)$ to say that x is *a* dog. The indefinite article is both retained and represented in our account, which is as it should be in a representation of our speech acts.

3. On the Unsaturated Nature of Concepts

Now what quantifier phrases stand for as noun phrases in the subject positions of our speech acts are referential concepts, and of course what predicate expressions stand for in such acts are predicable concepts. What unifies the speech act, and the underlying mental act, is the joint exercise of a referential and a predicable concept. As intersubjectively realizable cognitive capacities, these concepts have an unsaturated nature each complementary to the other, so that when exercised each saturates the other in a kind of mental chemistry resulting in a speech or mental act. The mutual saturation corresponds to what Gottlob Frege described as a first-level concept *falling within* a second-level concept, which is appropriately represented by the adjunction of a quantifier phrase with a predicate expression. We note that even though *falling within* is not same as *falling under*, i.e., where an object is said to fall under a first-level concept, nevertheless, according to Frege, both an object falling under a first-level concept, and a first-level concept falling within a second-level concept result in a truth value. That, however, is not how we understand the unsaturated nature of referential and predicable concepts.

In particular, as cognitive capacities, a referential and predicable concept, when exercised together in a speech or mental act, result in an event, not a truth value. The event may be just a mental event, e.g., a judgment, if it is not overtly expressed, or a speech act event as well if it is overtly expressed. The exercise of a referential concept informs the event with a referential, directed nature (or what Brentano and Husserl called a *presentation*[18]), and the exercise of a predicable concept informs the event with a predicable nature.

The unsaturated nature of referential and predicable concepts as cognitive capacities is radically different from the unsaturated nature of Frege's concepts. Frege's notion of a concept is that of an abstract function from objects to truth values, which means that it is nothing at all like a cognitive capacity. The unsaturated nature of concepts as cognitive capacities, on the other hand, is analogous to the nature of dispositional properties that real, non-abstract things might have, except that dispositions have a "were-would" nature, whereas cognitive capacities have a "were-could" nature. That is, if something has a dispositional property (such as solubility in water), then if it *were* in a certain context or placed under certain circumstances, then that thing *would* have a related but different property (such as being dissolved in water), whereas if a speaker *were* in an appropriate linguistic or mental context to apply a given concept, then that speaker *could*, but might not, exercise that concept in that context. This quasi-dispositional nature of concepts as cognitive capacities explains how concepts, unlike the momentary image-concepts of the medieval *suppositio* theory, can continue to exist as capacities even when they are not being exercised – just as dispositional properties somehow exist in nature even when they are not being manifested. It also explains how the same concept as an intersubjectively realizable rule-following cognitive capacity can be possessed, and exercised, by different people at the same time, as well as by the same person at different times. The unsaturated nature of concepts as cognitive capacities is a natural property that concepts have, and therefore, unlike Frege's notion, it is not part of an abstract ontology. Also, as the cognitive capacities underlying our rule-following abilities in the use of referential and predicable expressions, concepts have a certain kind of functionality different from Frege's abstract notion, specifically in how they are exercised and function in speech and thought. This kind of functionality explains their complementarity, i.e., how they mutually saturate each other in thought and speech.[19]

The exercise of a referential concept informs a speech or mental act (an event) with a referential nature, just as the predicable concept jointly exercised with

---

18 For a formal representation of Brentano's notion of a presentation in terms of the present logic of names see Cocchiarella (2013).
19 We do not claim that the unsaturated nature of concepts as cognitive capacities are irreducible to neurophysiological properties of the brain. As with the dispositional properties of physical objects, the issue of reducibility to occurrent states or properties remains controversial.

that referential concept informs that act with a predicable nature. Thus, every affirmative assertion that is syntactically analyzable in terms of a subject phrase and a predicate phrase (regardless of the complexity of either) is semantically analyzable in terms of an overt exercise of a referential concept with a predicable concept – and the assertion itself, as an event, is just the mutual saturation of their complementary structures in that speech act.

4. Nominalization and Relational Predication

We now extend our free first-order logic of names to a second-order predicate logic in which predicate expressions can be nominalized and occur as arguments of predicates, including of themselves. Given certain constraints that we will not go into here, Russell's paradox is not forthcoming, and the result is provably equivalent to the theory of simple types.[20] Predicate variables and quantifiers binding such are of course part of the logic. We will use the same symbol for the nominalized form of a predicate variable or constant. But when a predicate variable or constant occurs in its predicative role (along with one or more arguments), then it must have a pair of accompanying parentheses, as in $F(x)$ or $G(z, y)$, whereas the parentheses are deleted when the variable or constant occurs as a term or argument of a predicate, as, e.g., in $G(z, F)$, where G occurs as a predicate with its parentheses and F occurs as a term without parentheses.[21] We use the variable-binding λ-operator to represent complex predicates, as in the Russell predicate, $[\lambda x(\exists F)(F = x \land \neg F(x))]$, which is read as 'to be a concept that does not fall under itself', and which *qua* predicate expression stands for a predicable concept, i.e., a value of the bound (one-place) predicate variables, but which when nominalized does not denote a value of the bound first-order variables of our free first-order logic, because otherwise the paradox would be generated. In reasoning and trying to derive a contradiction on the basis of the Russell predicate we must know how to use the predicate, which means that the predicable concept for which it stands must exist as a cognitive capacity, otherwise we would not be able to reason with it. It does not mean, on the other hand, that the intensional content of the concept can be "objectified", i.e., exist as a value of the bound first-order variables.

What nominalized predicate expressions denote (when they in fact denote) are the intensions (or intensional contents) of the concepts (cognitive capacities) they stand for in their role as predicates. These intensions can be informally identified

---

20 See chapters 5 and 6 of Cocchiarella (1986), or §4.5 of Cocchiarella of (2007).
21 Our use of 'term' here should not be confused with what medieval logicians called terms.

with the functions from possible contexts of use of the predicates to the extensions of the concepts the predicates stand for in those contexts. They cannot of course be the concepts themselves, because as unsaturated cognitive concepts, the latter cannot be taken as values of the bound first-order variables. When a nominalized predicate occurs in a formula, we say that the concept the predicate stands for as a cognitive capacity has been *deactivated* (with respect to the nominalized occurrence of the predicate). It is deactivated of course because it is no longer functioning in that occurrence as a predicate.

One important use of nominalized predicates is for their occurrence as direct objects of intensional verbs, as in 'Jim wants to be president', which we can symbolize as:

($\exists x Jim$)$Wants(x, [\lambda y President(y)])$,

where the nominalized predicate $[\lambda y President(y)]$ is read as the infinitive phrase 'to be president'. We should note that the above predication, as well as all of the predications based on a relation discussed in this chapter, could (and perhaps should) be formulated in terms of a (complex) monadic predication as a λ-abstract. That would make it clear that what appears to be a relational predication is really a complex form of monadic predication, i.e., where predication is explained in terms of the mutual saturation of a referential concept and a monadic predicable concept. Thus, e.g., in the case now being discussed, we could (and perhaps should) symbolize the above sentence as:

($\exists x Jim$)$[\lambda x Wants(x, [\lambda y President(y)])](x)$.

We avoid doing so here mainly for simplicity and convenience of expression. Our point remains that predication is the mutual saturation of a referential and a (monadic) predicable concept even when the predicate expression is a complex based on a relation. We should note here, incidentally, that as a cognitive capacity a predicable concept, and similarly a referential concept, is not assumed to have a complex structure corresponding to the predicate, or referential expression, whose use it underlies – as might well be the case in a realist theory of predication about properties and properties of properties.[22]

In the above example we are not using 'is president' as a predicate. The predicate has been nominalized, indicating that the concept it stands for has been deactivated. Our view that a nominalized predicate occurring as direct

---

22 Properties of properties are what quantifier phrases stand for in a realist theory of predication.

(or indirect) object of a relational verb in a speech act denotes the intension of that predicate is similar to Richard Montague's view that such an occurrence is to be taken as denoting the sense of the predicate.[23] Montague gave this sort of analysis for the direct objects of all relations, whether or not those relations were intensional or extensional. That was an important insight especially with respect to how we should analyze direct (or indirect) objects that contain quantifier phrases. We will adopt the same strategy here for speech or mental acts, except that, instead of denoting the sense of a predicate occurring as the direct (or indirect) object, we take the occurrence of the predicate to be deactivated, which means that it is nominalized and denotes its intension instead.[24]

Thus, consider Montague's example of 'Jim wants to find a unicorn', which is symbolized as:

$(\exists x Jim) Wants(x, [\lambda y(\exists z Unicorn) Finds(y, z)])$.

Here the nominalized (complex) predicate $[\lambda y(\exists z Unicorn) Finds(y, z)]$ is read as 'to be a y such that y finds a unicorn', and here again on our view the predicate 'finds a unicorn' has been deactivated and is not being used to predicate finding a unicorn of anyone. The phrase 'a unicorn' occurs in this context as a quantifier phrase, but because it occurs within a deactivated predicate it is then understood as being itself deactivated, and hence as representing a deactivated referential concept. In other words, there is no reference to a unicorn in this example.

Now we do sometimes represent a quantifier phrase as being directly and not indirectly deactivated and hence occurring as a term of a complex predicate. The example Montague considered for his sense-denotation type theory was 'John seeks a unicorn', where the quantifier phrase 'a unicorn' is the direct object of 'seeks'. Instead of dealing with the sense of the quantifier phrase 'a unicorn' as in Montague's theory, we nominalize or deactivate it by first correlating the phrase with a predicate as follows:

$[\exists x Unicorn] =_{df} [\lambda y(\exists G)(y = G \land (\exists x Unicorn)G(x))]$,

---

23 See Montague (1970).
24 The difference between senses and intensions can be important, as in Montague's theory, or not, as in our theory, where much depends on how fine-grained an analysis is given to intensions. In our theory much depends on how we understand a context of use of language.

and then we nominalize or deactivate the predicate.[25] Here the predicate in question can be read as 'to be a concept under which a unicorn falls'. The intension of this predicate is what we take as the intension of the quantifier phrase 'a unicorn'. The sentence 'John seeks a unicorn' can then be symbolized as follows:

$(\exists yJohn)Seeks(y, [\exists xUnicorn])$.

A similar analysis applies to 'John finds a unicorn' as well:

$(\exists yJohn)Finds(y, [\exists xUnicorn])$,

which means that both sentences are analyzed as having the same logical form. In both cases we have interpreted the quantifier phrase 'a unicorn' as being nominalized, in which case we say that the referential concept it stands for has been deactivated, i.e., we are not here actively referring to a unicorn.

Now although these two sentences have the same logical form, there is a difference in the content of the predicate in each. In particular, whereas the predicate *Find* is extensional, the predicate *Seek* is not. Thus, given the extensionality of *Find*, we have as a meaning postulate:

$[\lambda yFinds(y, [\exists xA])] = [\lambda y(\exists xA)Finds(y, x)]$,

where A is a name variable. What the meaning postulate stipulates here is that 'to be a y such that y finds an A' is 'to be a y such that some A is such that y finds it'. From this meaning postulate it follows that there is a unicorn that John finds.[26] The same cannot be said for *Seek*, on the other hand, because *Seek* is an intensional and not an extensional verb. In this way we are able to give the same sort analysis that Montague gave in his treatment of quantifier phrases in English, but without having to resort to Montague's sense-denotation type theory.

---

25 The generalized schema for a "nominalized" quantifier phrase is:

   $[QxA] =_{df} [\lambda y(\exists G)(y = G \wedge (QxA)G(x))]$,

   where A is a name variable and Q is a quantifier. Note that when nominalizing a quantifier phrase we replace the parentheses that normally occur as part of the phrase with brackets.

26 This meaning postulate is of course an instance of the more schematic form:

   $[\lambda yFinds(y, [QxA])] = [\lambda y(QxA)Finds(y, x)]$,

   which can be applied to 'finds a few books', 'finds most books', etc.

5. Nominalization with Formula Operators

Now there is a problem with both Montague's and our analysis of a (complex) predication based on a relation with a quantifier phrase as direct object. Consider, e.g., the sentence 'Jim bought and ate an apple', the truth conditions of which can be symbolized as:

$(\exists x Jim)(\exists y Apple)(Bought(x, y) \land Ate(x, y))$.

As a speech or mental act, however, this won't do because of the double quantifier phrase. That is, as a speech or mental act what we have on this analysis are two active referential concepts, whereas in fact there is only one active reference in the assertion, namely to Jim.

We could try the following analysis in which the quantifier phrase 'an apple' has been nominalized and therefore deactivated:

$(\exists x Jim)[\lambda x[\lambda y(Bought(x, y) \land Ate(x, y))]([\exists y Apple])](x)$.

This analysis will not do as well, however, because by λ-conversion and the extensionality of *Bought* and *Ate*, it is equivalent to 'Jim bought an apple and Jim ate an apple', which is not equivalent to the original sentence, because on this analysis the apple he bought might be different from the apple he ate.

I have made several proposals elsewhere as to how to answer this problem in our conceptualist logic.[27] We will not review those proposals here, but I want to suggest a new, different proposal that is based on an issue regarding how to interpret tense and modal operators – and probably formula operators in general – in our speech or mental acts. I think this proposal works best for the example we are considering here, at least for our conceptualist logic even if not also for Montague's sense-denotation type-theoretical view.

Consider, for example, the denial that the round square is round, which, using the negation sign for the formula operator 'it is not the case', we can symbolize as

$\neg(\exists_1 x Square/Round(x))Round(x)$.

Here, of course, we are not referring to a square that is round. Nor of course are we predicating of the round square that it is not round (as Alexis Meinong would have it). Rather we are *denying* or rejecting the proposition that the round square is round. One way to read this is as 'That the round square is round *is not the case*', i.e., to understand the operator as a predicate, so that the above can also be symbolized as:

---

27 See Cocchiarella (2007, chapter 7.8).

$Not([\exists_1 xSquare/Round(x))Round(x)])$,

where the bracketed formula is read as the nominalized form of the indicated sentence and *Not* is taken as a predicate. Here we represent the nominalization of a formula by bracketing it. Predication, needless to say, is not activated in a nominalized sentence. Nor of course is reference, i.e., there is no active reference to a round square in this denial.[28] What a nominalized sentence denotes in our conceptualist logic is the intension of the sentence.

Now the point to our proposal is that the same kind of analysis applies to tense and modal operators, and in particular to the operator P for 'it was the case', so that 'It was the case that φ', where φ is a formula can be symbolized as $P\varphi$, but understood as $Was([\varphi])$, i.e., as 'that φ *was the case*'. Thus, when we assert the sentence 'That there is a king of France was the case', we are not actively referring to a king of France; rather we are only asserting that the proposition in question was the case or true in the past.

Why bring in the past-tense operator here when tense operators and their logic have been left as implicit in the background logic? The answer is because the predicate 'bought and ate' is in the past tense and really contains an implicit tense operator such as in 'x bought an apple *and then* ate it', which can be symbolized as:

$P(\exists yApple)(PBuys(x, y) \wedge Eats(x, y))$,

where $P(\exists yA)(P\varphi \wedge \psi)$ can be read as 'That $(\exists yA)(\varphi$ and then $\psi)$ was the case'. The 'and-then', in other words, can be defined in terms of the past-tense operator as indicated.[29] Notice that because the apple was eaten and hence no longer exists, we are not here (now) referring to an apple. The formula following the tense operator is to be interpreted as deactivated, in other words. The original sentence 'Jim bought and ate an apple' can now be symbolized as:

$(\exists xJim)[\lambda x\, P(\exists yApple)(\, PBuys(x, y) \wedge Eats(x, y))](x)$,

and although the quantifier phrase 'an apple' has not been nominalized in this formula, nevertheless, because of the past-tense context, the referential concept it stands for has been deactivated.

---

28 If we were to use this notation, we would need to add a meaning postulate connecting it to the standard notation so that we can engage in the usual deductions in the standard way. For convenience, we choose not to bother with introducing the new notation.

29 See Prior (1967:182). Prior did not use the 'and-then' operator for the purpose we are proposing here, however.

6. The Identity (Two-Name) Theory of the Copula

The idea that the direct (or indirect) object of a relational predicate is deactivated in a speech or mental act applies no less so to the copula in its use as a relational predicate than it does to transitive verbs such as 'seek' and 'find'. In addition, this application to the copula is quite useful in our analysis of predication.

Now the medieval terminist logicians and linguists interpreted the copula as an identity, a view that most modern philosophers and linguists reject, but that is because the terminist logicians took the copula to be no less than predication itself (for categorical sentences). On our view, however, taking the copula as predication is no more correct than taking any relation (such as 'seek' or 'find') as a form of predication, which does not mean that it cannot be the relational basis of a complex monadic predication.

We will use the expression '*Is*' to represent the copula as a two-place predicate, and we note that, like the transitive verb 'find', the copula *Is* is extensional. Our meaning postulate in this regard transforms the copula into a strict identity. With a nominalized quantifier phrase as the direct object of the copula, the schematic meaning postulate for *Is* is as follows:

$$[\lambda x Is(x, [\exists y A])] = [\lambda x (\exists y A)(x = y)],$$

where A is a name variable. Thus, an assertion of the sentence 'Jan is a teacher' can now be symbolized as:

$$(\exists x Jan) Is(x, [\exists y Teacher]),$$

which, by our meaning postulate regarding the extensionality of the copula, is equivalent to:

$$(\exists x Jan)(\exists y Teacher)(x = y),$$

which is a more perspicuous representation of its truth conditions. It is important to keep in mind here that the predicate in the speech act that Jan is a teacher is $[\lambda x Is(x, [\exists y Teacher])]$, and not the copula or the identity sign itself. In other words, it is the predicable concept that this predicate stands for that is exercised in the speech act in question (and mutually saturated by the referential concept regarding Jan), and not the relational (identity) concept that *Is* stands for. As already noted, we have avoided writing out the λ-abstract throughout only for convenience.

It was something like the above sort of analysis that came to be called the two-name, or identity, theory of the copula in terminist logic. Apparently, Ockham and other terminists thought that every affirmative categorical proposition amounted to asserting an identity between the personal suppositions of the subject and predicate terms of the proposition, as, e.g., the suppositions of the names 'Jan' and 'teacher' in an assertion of 'Jan is a teacher'. A negative judgment was construed as a denial of such an identity. As a result, the identity theory of the copula came to be developed as a theory of the truth conditions of categorical propositions, a theory that is now referred to as *the doctrine of supposition proper*.[30]

Consider, for example, an assertion of 'Every whale is a mammal'. On our analysis this sentence has the logical form,

$(\forall x Whale)Is(x, [\exists y Mammal])$,

which, by the meaning postulate for *Is*, reduces to:

$(\forall x Whale)(\exists y Mammal)(x = y)$.

What this means in terminist logic is that *each* supposition of the categorical term 'whale' and *some* supposition of the categorical term 'mammal' are identical.

The terminists also understood categorical propositions with a predicate adjective as being based on the 'is' of identity as well. Thus although an assertion of 'Every raven is black' is represented in our logic by

$(\forall x Raven)Black(x)$,

this does not mean (as some philosophers have thought) that the terminists interpreted the sentence as an identity between ravens and black, or blackness, or separate blacknesses for different ravens. Rather, the predicate adjective 'black' was interpreted as an attributive adjective, so that to say a thing is black is to say that it is a black thing.[31] Predicate adjectives, in other words, were analyzed by the terminists as attributive adjectives applied to the common name 'thing', which is the opposite of our view that attributive adjectives are to be analyzed as predicate adjectives in relative clauses. But, as already noted, the common name 'black thing' is equivalent to the complex common name 'thing that is black', which is symbolized in our system as '*Thing/Black(x)*'. Thus, whereas the terminist logician would interpret 'Every raven is black' as 'Every raven is a black thing', we can represent the terminists' analysis as 'Every raven is a thing that is black' as follows:

$(\forall x Raven)Is(x, [\exists y Thing/Black(y)])$,

which, by our meaning postulate for *Is*, is equivalent to

---

30 See Scott (1966a: 30).
31 See Normore (1985: 194).

$(\forall x Raven)(\exists y Thing/Black(y))(x = y)$,

which, by our meaning postulates for the relative clause operator, reduces in logic to the above formula with 'black' as a predicate adjective. What this shows is that we have not really escaped the predication of 'is black' by 'thing that is black', since the occurrence of 'is black' now occurs in the relative clause, and a repeat of the analysis would only go on in an infinite regress to 'thing that is a thing that is ... a thing that is ...'. In the end we must ground the predication in terms of the mutual saturation of a referential concept such as is expressed by 'Every raven' and a predicable concept such as is expressed by 'is black'.

Negative categorical sentences such as 'No whale is a fish' were interpreted by the terminists as denials or negations. That is, to assert that no whale is a fish is to deny that some whale is a fish:

$\neg(\exists x Whale)Is(x, [\exists y Fish])$,

which, by the meaning postulate for Is reduces to

$\neg(\exists x Whale)(\exists y Fish)Is(x, y)$,

and which for the terminists amounted to a denial that some supposition of the name 'whale' is identical with a supposition of the name 'fish'.

Finally, let us turn to the logical form of a negative particular categorical sentence such as 'Some raven is not black'. This sentence can be symbolized with a λ-abstract as follows:

$(\exists x Raven)[\lambda x \neg Black(x)](x)$.

Our use of the λ-abstract here is so as to emphasize that the negation is an internal part of the predicate. This sentence would be understood in terminist logic as the statement that 'some raven is not a black thing', which is symbolized as

$(\exists x Raven)[\lambda x \neg Is(x, [\exists y Thing/Black(x)])](x)$.

This last formula, by λ-conversion and the meaning postulate for Is, is equivalent to

$(\exists x Raven)(\forall y Thing/Black(x))(x \neq y)$,

the truth conditions for which are that some supposition of the common name 'raven' is not identical with any supposition of the complex common name 'thing that is black'.

We will not go on with further examples of the terminists' two-name or identity theory of the copula.[32] It is important to note here, however, that the idea of the copula as an identity is not wrong, but it needs to be seen as restricted to contexts with a quantifier phrase occurring as the direct object of the copula, and even then, it is to be understood as the basis of a complex relational predicate. This applies even when the copula is between two proper names, as in 'Cicero is Tully', which can symbolized as:

$(\exists x Cicero)[\lambda x Is(x, [\exists y Tully])](x)$,

but which – once names can be transformed into terms (in the extended logic that is to follow) – is equivalent to:

Cicero = Tully.

The important thing is to distinguish here the difference between the logical form that represents the speech act as based on a referential and predicable concept and the logical form that perspicuously represents its deductive role and hence its truth conditions (which in this case is the deductive logic of identity).

7. Plural Reference and Predication

We now turn to an analysis of plural reference and predication. As we will see, our account of the unity of predication in a speech or mental act applies no less so to plural reference and predication – and also to mass noun reference and predication – than to singular reference and predication. Both plural and singular predication, we maintain, are categorially on a par, i.e., there is no category difference or difference of level in English between singular and plural predication the way there is between subjects and predicates, though that is not how some philosophers represent the situation.[33] Plural reference will of course be to pluralities, and pluralities will be what plural predication is about. What we need accordingly are logical forms that represent pluralities, and along with such forms a logic of the truth conditions of our references to, and predications of, pluralities.

Now it should be noted that our treatment of names, proper or common, is not restricted to count nouns. That is so because the logic we have developed for count nouns applies to mass nouns as well. One important move in developing a logic for both plurals and mass nouns is to allow the common names that have so

---

32 See Cocchiarella (2001) for more examples.
33 Boolos (1986), e.g., takes second-order predicate logic to be the logic of plurals, and others such as MacKay (2006) take it to be a totally new syntactic category.

far occurred exclusively as parts of quantifier phrases to occur also as "terms" or arguments of predicates, i.e., as denoting expressions on a par with the variables of first-order logic. That of course is exactly how proper names in particular are normally represented in standard logic; but our point is that the same is to be done with common names as well (which is essentially the way they are understood in Leśniewski's logic of names). And when such a common name is a count noun and occurs as an argument of a predicate then it is understood to denote a plurality (unless it denotes nothing).

In other words, we will allow for the nominalization or transformation of all names, proper or common, into "terms", i.e., expressions that can occur as arguments of predicates (the way the variables of first-order logic do).[34] But for now we will restrict our attention to count nouns. When a count noun such as 'man' is nominalized in the manner indicated, what we get in English is the plural 'men' or the noun 'mankind' as when we say that Socrates is a member of mankind, not meaning that Socrates is a member of the set of people, but only that he is one among men (or people). Being one *among*, or being a *member of*, a collective or group such as mankind does not mean being a member of a set. Membership has a meaning or use in contexts about pluralities as collectives other than sets, and it is that use that we are concerned with here. In addition to "nominalized" names, proper or common, conjunctions of names, as in 'George, Harry and Jim', are also taken to denote pluralities, or collective groups, as for example in the statement that Russell and Whitehead are the coauthors of *Principia Mathematica*, or that George, Harry, and Jim are playing cards (together).

Now because names, proper or common, and complex or simple, can be transformed into arguments of predicates, i.e., into terms, we need to add to our logic a variable-binding operator that generates complex names the way that the λ-operator generates complex predicates. We will use the cap-notation with brackets, $[\hat{x}A/\ldots x\ldots]$, for this purpose. Accordingly, where A is a name, proper or common and complex or simple, we take $[\hat{x}A]$ to be a complex name, but one in which the variable x is bound. Thus where A is a name and φ is a formula, $[\hat{x}A]$, $[\hat{x}A/\varphi]$, and $[\hat{x}/\varphi]$ are names in which all of the free occurrences of x are bound. We read these expressions as:

$[\hat{x}A]$ is read as '(the) A things',

---

34 To be a term in this logic is not the same as to be a singular term, i.e., a term that denotes at most one single object or thing. Pluralities in particular will be denoted by terms and so will the various parts of what mass nouns denote. By a term we mean only an expression that can occur as an argument of predicates.

$[\hat{x}A/\varphi]$ is read as '(the) A things that are φ', and $[\hat{x}/\varphi]$ is read as '(the) things that are φ'.

Thus for example, the conjunction of 'George, Harry, and Jim' can be represented by $[\hat{x}/x = George \lor x = Harry \lor x = Jim]$, which is read 'the things that are either George, Harry, or Jim'. Usually, though, it is just read as a conjunction of names, as in 'George, Harry and Jim are playing poker'.

Now we assume that in a given speech context we can distinguish the common names that represent count nouns from those that represent mass nouns. We do not assume, on the other hand, that there is an absolute, fixed distinction between count and mass nouns, but only that we can make such a distinction in particular contexts of use of language.[35] Plural reference in English, of course, involves only count nouns.

In regard to the logical forms for plural reference and predication, we extend the inductive definition of the meaningful (well-formed) expressions of our conceptualist logic to include the following clauses:

(1) if A represents a common count noun (in a given context), then A^P is a *plural name* (in that context);
(2) if A represents a common count noun (in a given context), x is a first-order variable, and φx is a formula, then $[\hat{x}A/\varphi x]^P$ and $[\hat{x}/\varphi x]^P$ are *plural names* (in that context);
(3) if $A/\varphi(x)$ represents a (complex) common count noun (in a given context), then $(A/\varphi x)^P$ is $A^P/[\lambda x \varphi x]^P(x)$ and $[\hat{x}A/\varphi x]^P$ is $[\hat{x}A^P/[\lambda x \varphi x]^P(x)]$;
(4) if F is a one-place predicate constant, or of the form $[\lambda x \varphi(x)]$, then F^P is a one-place *plural predicate constant*; and
(5) if A^P is a plural name (in a given context), x is a first-order variable, and φ is a formula, then $(\forall x A^P)\varphi$ and $(\exists x A^P)\varphi$ are formulas.

We read '$(\forall x Man^P)$' as the plural phrase 'all men' and '$(\exists x Man^P)$' as the plural phrase 'some men'. In other words we take these phrases as representing plural referential concepts. Similarly, we read the complex predicate $[\lambda x(\exists y Man)(x = y)]^P$ as the plural predicate 'men'. We note that a *plural* name is not a name *simpliciter*, and hence a plural name does not occur as a term or argument of predicates. That role is already filled by nominalized names *simpliciter*. That is the main difference here between plural names and names *simpliciter*: plural names do not occur as terms in our basic logical forms, which means that plural names occur only in

---

35 See, e.g., Pelletier (1975: 456), regarding the notion of a "universal grinder" that can change a count noun into a mass noun in a given context.

the logical forms that represent our speech or mental acts, and in particular as plural predicates in plural predication or as part of plural quantifier phrases in plural reference. Finally, it is important to note that we use meaning postulates to connect the logical forms of our speech or mental acts with the logical forms that give a more perspicuous representation of the deductive implications of our speech or mental acts, and hence of their truth conditions.

We note that only monadic predicates are pluralized. A two-place relation R can be pluralized in either its first- or second-argument position, or even in both, by using a λ-abstract, as, e.g.,

$[\lambda x R(x,y)]^P$,
$[\lambda y R(x,y)]^P$,
$[\lambda x[\lambda y[R(x,y)]^P(y)](x)]^P$,

respectively; and a similar observation applies to n-place predicates for $n > 2$.

In turning to examples of plural reference and predication, let us consider the sentence 'Whales are mammals', where of course by 'whales' we mean 'all whales'. As a thought or speech act, the sentence can be symbolized as follows:

$(\forall x Whale^P) Mammal^P(x)$.

This formula of course should be equivalent to 'Every whale is a mammal', a statement that involves only singular predication. In fact, the equivalence is provable in the logic we describe below. It is based on the same logical considerations given for 'All men are mortal' being equivalent to 'Every man is mortal' described below.

For an example that is not reducible to singular predication, consider the statement that some men are playing poker (together), which we can symbolize as follows:

$(\exists x Man^P) Playing\text{-}poker^P(x)$.

Playing-poker is interpreted here as plural because the men are playing poker together. To be sure, if some men *are* playing poker (together), then some man, e.g., George, *is* playing poker, where 'playing poker' is now in the singular. But playing poker together, i.e., where 'playing poker' is pluralized, is not a distributive predicate, which means that it cannot be reduced to, or completely analyzed in terms of, 'playing poker' in the singular.

The question now is how are we to understand the truth conditions of a statement of the above form, or more generally of the form $(\exists x A^P) F^P(x)$, and also of the form $(\forall x A^P) F^P(x)$ as well? It is one thing to say that pluralities are the values of the variables in a formula, but what are pluralities, and how are they to be represented here both as to what we are referring to and what the predicate says about them? What, in other words, is the semantics or logic of pluralities?

Our answer is that a plurality is essentially what Bertrand Russell called a class as many as a plurality in his 1903 *Principles of Mathematics*, by which Russell did not mean either a set or a class as an abstract entity.[36] To be sure, Russell gave up the notion of a class as many as a plurality after 1903 as hopelessly confused, and that is just what Peter Geach claimed when he described it as "radically incoherent".[37] But Geach was wrong, because by following Russell s three basic principles regarding classes as many we have been able to formalize a logic of classes as many that is not only coherent but consistent as well.[38]

Russell assumed three important features about his notion of a plurality as a class as many, and, as indicated, each is valid in our logic of classes as many. The first is that a vacuous common count noun, i.e., a common name that names nothing, denotes nothing, which is not the same as having an empty class as its extension. Thus, according to Russell, "there is no such thing as the null class [in the sense of a class as many as a plurality], though there are null class-concepts", i.e., common-name concepts that have no extension, or denote nothing.[39] Thus, for example, the plurality, or class as many, of things that are not self-identical, $[\hat{x}/x \neq x]$, denotes nothing, i.e., $\neg(\exists y)(y = [\hat{x}/x \neq x])$ is a valid (provable) thesis of our logic of plurals, which it will be remembered is based on a free first-order logic. We use Λ as the symbol for the empty class as many, i.e.,

$$\Lambda =_{df} [\hat{x}/x \neq x].$$

As indicated, it is provable in our logic of classes as many as pluralities that there is no empty class as many, i.e.,

$$\neg(\exists x)(x = \Lambda)$$

is provable in our logic.

Now where membership in a plurality is defined as follows:

$$x \in y \leftrightarrow (\exists A)[y = A \wedge (\exists zA)(x = z)],$$

the Russell plurality, or class as many of things that are not a member of themselves, also does not exist, i.e.,

---

36 For a 17th century anticipation by Thomas Vincentius Tosca of the distinction between a class as many and a class as one see Angelelli (1979).
37 Geach (1980a: 225).
38 See, e.g., Cocchiarella (2002).
39 Cocchiarella (2002, §69).

$\neg(\exists x)(x = [\hat{y}/(\exists A)(y = A \wedge (y \notin A))])$

is also a provable thesis of logic of plurals. Note, incidentally, that in the definition of ∈ the occurrence of A in $(y = A)$ is as a term denoting the plurality of things that fall under the name concept that A stands for, whereas the occurrence of A in the quantifier phrase $(\exists zA)$ stands for the name concept itself.

The second important feature assumed by Russell is that the class as many as a plurality that is the extension of a count noun that names just one thing is just that one thing. In other words, unlike the singleton sets of set theory, which are not identical with their single member, the class as many that is the extension of a common count noun that names just one thing is none other than that one thing. This is not really odd in fact because we sometimes refer to, or speak of, e.g., 'some people' in a context when, it turns out, there is just one person involved, just as we also occasionally use 'some person' in a context when there is more than one person involved. In addition, overall the logic of pluralities comes out much more smoothly when we allow that a plurality might consist of just one thing. Thus given that 'Socrates' denotes one thing, then

$(Socrates \in Socrates) \wedge (\forall x/x \in Socrates)(x = Socrates)$

is a true sentence, and similarly so is

$(\exists xSocrates)(\forall ySocrates)(x = y)$.

Finally, the third feature cited by Russell is that, unlike sets, classes as many are literally made up of their members, which is why they are also called pluralities (*Vielheiten*), and not things that can themselves be members of classes.[40] Thus, according to Russell, "though terms may be said to belong to ... [a] class [as many], the class [as a plurality] must not be treated as itself a single logical subject".[41] Thus, if a plurality or class as many is made up of at least two objects, then that plurality cannot itself be a member of a class as many, i.e., it cannot be one among a plurality.

In turning to the semantics or logic of classes as many as pluralities, we note that inclusion in a class as many, or being part of a plurality, can be defined in terms of membership, i.e., the relation of being among:

---

40 The idea that sets also have their being in their members is philosophically problematic when we consider the empty set, which would mean that it has its being in nothing, and hence that it itself is nothing. But the empty set is essential to pure set theory and therefore is not nothing.
41 Russell (1903, §70).

$x \subseteq y \leftrightarrow (\forall z)[z \in x \rightarrow z \in y]$.

As indicated, the inclusion relation, $\subseteq$, of our logic of classes as many can also be used to represent the part-to-whole relation that is essential to any account of mass-noun reference and predication. The logic of mereology as described in the Leonard-Goodman's calculus of individuals is in fact reducible to our logic of classes as many, or at least the atomistic (free-logic) version is.[42] Proper inclusion, $x \subset y$, is definable of course in terms of inclusion:

$x \subset y \leftrightarrow x \subseteq y \wedge y \nsubseteq x$.

What is an "atom" (in the sense of our logic), or "single thing", i.e., an "individual", is analyzable in the logic of classes as many as follows:

$Atom =_{df} [\hat{x}/\neg(\exists y)(y \subset x)]$.

That is, to be an atom, or individual, is to be something that has no proper sub-part.[43] In the case of mass nouns, which we will deal with next, an atom is the same as a *minimal* part. There are a number of interesting features of the logic of classes as many as pluralities that we will not go into here, but the details of which can be found in my paper of 2002. We should also note, however, that combining pluralities of the same kind results in a plurality of that same kind:

$(\forall x/x \subseteq A)(\forall y/y \subseteq A)[(x \cup y) \subseteq A]$.

In other words, pluralities are cumulative. As we will see, so too are what mass nouns denote.

Now the semantics or logic underlying plural reference is given in terms of the logic of classes as many as pluralities.[44] As already noted, we use meaning postulates to connect the logical forms at the level of our speech and mental acts with the logical forms at the level of the underlying logic of classes as many. Thus, for the plural reference of 'Some A^P are ...', we have the following principle as a meaning postulate (for plurals):

(MPP1) $\qquad (\exists x A^P)\varphi(x) \leftrightarrow (\exists x/x \subseteq [\hat{y}A])\varphi(x)$.

---

42 See Eberle (1970, Chapter 2), for a reconstruction of the calculus of individuals in a free first-order logic.
43 This 'atom' terminology goes back to Nelson Goodman and the so-called Leonard-Goodman calculus of individuals.
44 For a set-theoretic semantics as well see the appendix of Cocchiarella (2002).

Accordingly, for the statement that some men are playing poker we have the following biconditional connecting the logical form of the speech act with the logical form of the truth conditions of that act:

$(\exists x Man^P) Playing\text{-}poker^P(x) \leftrightarrow (\exists x/x \subseteq [\hat{y}Man]) Playing\text{-}poker^P(x).$

Now given that 'playing poker' is a distributive predicate in only one direction, i.e., that

$Playing\text{-}poker^P(x) \rightarrow (\forall y/y \in x) Playing\text{-}poker(y)$

is true, then the right-hand side of the above biconditional is provably equivalent to the formula

$[\hat{y}\, Man/Playing\text{-}poker(y)] \neq \Lambda.$

That the plurality (class as many) of men playing poker is not empty is equivalent of course to saying that some men are playing poker.

Let us turn now to how the universal plural 'All^P', the meaning postulate of which is given as follows:

(MPP2) $(\forall x A^P)\varphi(x) \leftrightarrow (\forall x/x \subseteq [\hat{y}A])\varphi(x).$

Now given that the cognitive structure of an assertion of 'All men are mortal' is represented as,

$(\forall x Man^P) Mortal^P(x),$

then, by (MPP2), it follows that, semantically, the assertion amounts to predicating mortality to every plurality of men,

$(\forall x/x \subseteq [\hat{y}Man]) Mortal^P(x),$

which is equivalent to saying that the members of the entire group of men taken collectively *are* mortal:

$(\forall x/x = [\hat{y}Man]) Mortal^P(x).$

In conceptualist terms, this formula comes very close to what Russell claimed in his 1903 *Principles*, namely, (to use Russell's terminology) that the denoting phrase 'All men' in the sentence 'All men are mortal' denotes the class as many of men.

Note that the predicate 'mortal' is distributive (in both directions), a fact that is represented by the following meaning postulate:

$(\forall x)[Mortal^P(x) \leftrightarrow (\forall y/y \in x) Mortal(y)].$

It follows, accordingly, that the statement that all men *are* mortal,

$(\forall x Man^P) Mortal^P(x)$,

is provably equivalent to the different statement that every man *is* mortal:

$(\forall x Man) Mortal(x)$.

Of course, there are plural predicates that are not distributive (in either direction) and which therefore cannot be reduced in this way.

Plural identity, as in 'The triangles that have equal sides *are* (identical with) the triangles that have equal angles', is based on the plural of the copula for both subject and direct object. The relation *Are* as the plural of *Is* can be represented by the following λ-abstract:

$Are =_{df} [\lambda x[\lambda y Is(x,y)]^P]^P$.

For convenience, we will ignore going through several applications of the meaning postulate about *Is* that leads to a reduction of *Are* to an identity between terms. We will simply use the identity sign instead. In other words, although 'A^P are B^P' is definable as $[\lambda x[\lambda y Is(x,B^P)]^P]^P(A^P)$, it is simpler for our present purpose here to represent 'A^P are B^P' simply as '$A = B$', with which, by meaning postulates, $[\lambda x[\lambda y Is(x,B^P)]^P]^P(A^P)$ is provably equivalent.

Plural identity, we note, is involved in the truth conditions for statements with two plural definite descriptions, just as singular identity is involved in the truth conditions for two singular definite descriptions. Thus, just as the statement that the spy is the bald man can be symbolized as follows:

$(\exists_1 x Spy) Is(x, [\exists_1 y Man/Bald(y)])$,

which reduces to:

$(\exists_1 x Spy)(\exists_1 y Man/Bald(y))(x = y)$,

so too in an entirely similar way we can formalize the statement that the spies are the bald men:

$(\exists_1 x Spy^P) Are(x, [\exists_1 y Man^P/Bald^P(y)])$,

which reduces to:

$(\exists_1 x Spy^P)(\exists_1 y Man^P/Bald^P(y))(x = y)$.

Now the logical analysis of a statement of the form The A^P are F^P can be given as follows:

$(\exists_1 x A^P) F^P(x) \leftrightarrow (\exists x A^P)[(\forall y A^P)(y = x) \wedge F^P(x)]$, for when the plural definite description is being used with existential presupposition. This of course is similar to the Russellian analysis for the singular form 'The A is F'. From this it follows that if F is distributive then the plural definite description 'The A^P are F^P' denotes the class as many of A that are F, i.e.,

$$(\exists_1 x A^P/F^P(x))(x = [\hat{y}A/F(y)]),$$

which explains why we read $[\hat{y}A/F(y)]$ as the 'A's that are F'.

For a sentence of the form 'The A's that are F are the B's that are G', as in 'The triangles that have equal sides *are* (identical with) the triangles that have equal angles', the analysis or symbolization can be given as follows:

$$(\exists_1 x A^P F^P(x)) Are(x, [\exists_1 y B^P/G^P(y)]),$$

which in our logic reduces to:

$$(\exists_1 x A^P F^P(x))(\exists_1 y B^P/G^P(y)])(x = y),$$

which of course can be further reduced by means of the above formula for plural definite descriptions.

Finally, we note that plurals also can be used with numerical quantifier phrases, as in 'There are twelve Apostles', as well as with numerical predicates, as in 'The Apostles *are* twelve'. As is well-known, the predicate 'twelve' can be defined in terms of numerical quantifier phrase, such as $(\exists^{12} y)$. For example, one analysis of 'The apostles are twelve' is the following:

$$(\exists_1 x Apostle^P)[\lambda x(\exists^{12} y)(y \in x)]^P(x).$$

The quantifier phrase $(\exists^{12} y)$ of course is readily definable in first-order logic (with identity).[45] We can also define the predicate 'twelve' more generally as the concept of pluralities that have twelve members:

$$12 =_{df} [\lambda x(\exists A)(x = A \wedge (\exists^{12} yA)(y \in x))].$$

A similar analysis applies of course for each natural number. 'The Apostles are twelve' can now be simply represented as:

$$(\exists_1 x Apostle^P)12(x).$$

---

[45] We note that given the extensionality of the membership predicate $\in$ (and its converse) the predicate $[\lambda x(\exists^{12} y)(y \in x)]$ is provably equivalent to $[\lambda x \breve{\in}(x, [\exists^{12} y])]$, where $\breve{\in}$ is the converse of $\in$. The active quantifier $(\exists^{12} y)$ in the predicate of the above formula can then be replaced by one that is deactivated for the speech act in question. We avoid doing so here for convenience.

The sentence 'There are twelve Apostles' is more involved, however, because of the presence of two seemingly active quantifier phrases, namely 'twelve' and 'There are'. An alternative reading of this sentence is 'Twelve Apostles there are', where the quantifier phrase 'there are' is now (part of) the predicate. Treating 'there are' as (part of) the predicate amounts to pluralizing 'there is' as a predicate, specifically as $[\lambda x(\exists y)(x = y)]^P$.[46] The sentence 'Twelve Apostles there are' can then be analyzed as:

$(\exists^{12} x Apostle^P)[\lambda x(\exists y)(x = y)]^P(x),$

which is reducible to:

$(\exists^{12} x Apostle)(x = x),$

which amounts to saying indirectly that there are twelve Apostles.[47] A similar analysis can be given more generally for sentences of the form 'There are n many A^P'.

We conclude that our conceptualist account of predication in terms of the mutual saturation of referential and predicable concepts as unsaturated cognitive capacities applies to plural predication as well as to singular predication. In addition we have shown how sentences based on plural reference and predication can be analyzed in terms of the logically perspicuous forms of our conceptualist logic, which includes a logic of classes as many as pluralities.

8. Mass Noun Reference and Predication

We now turn to our analysis of mass-noun reference and predication. The details of our analysis are similar to that for plural reference and predication, which is not surprising given the grammatical similarity between plurals and mass nouns, such as the fact that mass nouns can take only those determiners that can also be used with plurals.[48]

Mass nouns can be complex, such as 'polluted water', 'tall grass', 'modern furniture', etc., as well as simple, such as 'milk', 'gold', 'furniture', etc. Some nouns in

---

46   The predicate $[\lambda x(\exists y)(x = y)]^P$ can be replaced by $[\lambda x Is(x,[\exists y])]^P$, with which it is identical by meaning postulates, so that there is no active quantifier in the predicate.

47   Note that the consequent of the formula is equivalent to a trivially provable thesis, namely that whatever is in x is something:

$[\lambda x(\exists y)(x = y)]^P(x) \leftrightarrow (\forall z/z \in x)(\exists y)(z = y),$

which means that the consequent is equivalent to $x = x$.

48   For an account of the similarities between mass nouns and plurals, see, e.g., Nicolas (2008).

some contexts function as mass nouns and in other contexts as count nouns, e.g., 'chicken' as in 'George ate a lot of chicken' (mass noun) and 'George has five chickens in his back yard' (count noun).

As already noted for count nouns, we assume that we can distinguish in a given speech context the common names that represent mass nouns from those that represent count nouns; but we do not assume that there is an absolute, fixed distinction between count and mass nouns.[49]

In regard to the logical forms for expressing mass-noun reference and predication, we extend the inductive definition of the meaningful (well-formed) expressions of our conceptualist framework to include the following clauses:

(1) if A represents a mass noun (in a given context), then A^M is a *mass name* (in that context);

(2) if A represents a mass noun (in a given context), x is a first-order variable, and φx is a formula, then $[\hat{x}A/\varphi x]^M$ and $[\hat{x}/\varphi x]^M$ are (complex) *mass names* (in that context);

(3) if $A/\varphi(x)$ represents a (complex) mass noun (in a given context), then $(A/\varphi x)^M$ is $A^M/[\lambda x \varphi x]^M(x)$ and $[\hat{x}A/\varphi x]^M$ is $[\hat{x}A/^M[\lambda x \varphi x]^M(x)]$; and

(4) if A^M is a mass name (in a given context), x is a first-order variable, and φ is a formula, then $(\forall x A^M)\varphi$ and $(x A^M)\varphi$ are formulas.

We note that clause (3) is needed to allow for the representation of such complex mass nouns as 'polluted water', i.e., 'water that is polluted'. In addition, we note that in referring to all, some, a lot of, etc., water, we are referring to parts of water that have indefinitely many subparts that are again parts of water and that stand in the same relation to the larger part as the larger part to the whole. But having indefinitely many subparts, we note, is not the same as having infinitely many subparts; that is, it does not mean that there is an infinite descent of subparts.[50] The various parts of water are not discrete, well-delineated single objects, i.e., "individuals", in the way that the minimal parts are, namely, the molecules of water, which are discrete and well-delineated. Thus, although the minimal parts of what a mass noun denotes are individuals, and hence can be individuated, it is only the non-minimal parts that are problematic in this way.

We note that as with plural names *mass* names are not names *simpliciter*, and hence a mass name cannot occur as a term or argument of predicates. That role

---

49   We will not deal with abstract mass nouns in this paper.
50   Having indefinitely many subparts is different from having infinitely many subparts partly because subparts themselves might have subparts, and in general there is no principle way to count all of the subparts.

is already done by nominalized names *simpliciter*. That is the main difference here between *mass* names and names *simpliciter:* mass names do not occur as terms in our basic logical forms. In other words, mass names, like plural names, occur only in the logical forms that represent our speech or mental acts, and in particular as mass predicates in their role as mass predications or as part of mass quantifier phrases for mass reference. Finally, it is important to emphasize once again that we use meaning postulates to connect the logical forms of our speech or mental acts with the logical forms that give a more perspicuous representation of the deductive implications of our speech or mental acts, and hence of their truth conditions.

Now it is noteworthy that our underlying logic of classes as many contains an atomistic (free-logic) mereology as a subsystem, which of course is natural for pluralities – but not, according to some linguists and philosophers, for the semantics of mass nouns. Nevertheless, despite the negative view, we maintain that an atomistic mereology is appropriate for mass nouns. In our logic we can speak of the extension (at a given time) of a mass noun, which in fact consists just of the minimal parts of what that mass noun denotes (at that time). Of course that is not a problem for mass nouns representing elementary substances and chemical compounds, because such mass nouns in fact denote at a given time the total class as many of individual atoms or molecules of those substances and compounds existent at that time. The mass noun 'gold', for example, denotes now all of the gold, and therefore all of the gold atoms, that exists now; and 'water' similarly denotes now all of the water, and therefore all the molecules of water that exist now. Pieces or bits of gold and bodies of water, large or small, are all made up of the atoms of gold or molecules of water, and therefore they are really subpluralities – or subclasses of classes as many – of the total class as many, or plurality, of gold and water, respectively.

The mass nouns for space and time, to be sure, have traditionally been assumed not to have minimal parts, and of course mathematically that is a consistent assumption. In particular, it is not just conceivable, but consistent as well, that there should be an infinite descent of the parts of space and time. Nevertheless, it is noteworthy that in modern quantum physics space and time are "quantized", and therefore do have minimal parts. The "Planck length" of 10^{-33} cm. is the smallest length physically possible in quantum mechanics, and there is a smallest physically possible time as well, namely, the time it takes for light to cross the Planck length, which is 10^{-43} seconds. This means that real space and time are not infinitely divisible after all.

Mass nouns other than those for elementary substances and molecules built up from such elements are also unproblematic, we maintain. The mass noun 'furniture', for example, clearly denotes all of the individual pieces of furniture, and similarly 'jewelry' and 'silverware' denote all the individual pieces of jewelry or silverware. Mass nouns such as 'wine', 'milk', 'coffee', etc., also clearly do not have an infinite, unending descent of parts that are also wine, milk or coffee.

Is there an atomistic theory for the semantics of mass nouns implicit in natural language or our commonsense framework? That is certainly not obvious; but it is an interesting hypothesis. It has been suggested by Henry Laycock, for example, who thinks that "a kind of atomic theory is implicit in the ordinary use of mass terms based on our experience of the behavior of stuff".[51] In other words, according to Laycock, we should "think of stuff as a plurality of things, each of the same kind for any given kind of stuff", and in that way we can "construe any mass term 'm' as a plural sortal of the form 'm elements'".[52] An atomistic mereology for mass nouns is appropriate, in other words, not just because there is no infinite descent in nature, but also because such a view would explain the source of the grammatical similarities between the semantics of mass nouns and that of plurals. A similar position is taken by Gennaro Chierchia who suggests that a "mass noun simple denotes a set of ordinary individuals plus all the pluralities of such individuals".[53] Chierchia recognizes that this "view is an 'atomistic' one: we are committed to claiming that for each mass noun there are minimal objects of that kind, just like for count nouns, even if the size of these minimal parts may be vague".[54]

In any case, as we will see, our atomistic logic of classes as many as pluralities works just as well for mass nouns as it does for plurals. In addition, one consequence of this logic is that each nominalized mass noun denotes at any given time the entire class as many of its minimal parts, i.e., the "atomic", minimal parts of the mass noun that exist at that time, which also means that it denotes the sum of all of its parts (existent at that time).

---

51   Laycock (1972: 39).
52   Laycock (1972: 38).
53   Chierchia (1998: 54). Unlike Chierchia, however, I assume that a mass noun occurring as a term denotes not the *set* of its minimal parts but, *qua* plurality, the class as many of its minimal parts. Also, whereas Chierchia maintains that every plurality consists of two or more members, I allow single individuals to be pluralities. (I did initially require groups to contain two objects in my 2002 paper, but I later changed that in my 2009 paper.)
54   Chierchia (1998: 54).

The meaning postulates connecting the logical forms representing our speech and mental acts in the case of mass nouns are given in a way entirely analogous to those already stipulated for plural reference and predication. Thus, just as the plural count noun phrase 'some A^P', in symbols, $(\exists xA^P)$, was interpreted as $(\exists xA/x \subseteq A)$, which can be read as 'Some plurality of A things', so too a similar interpretation applies to mass nouns. That is, where A^M is taken as a mass name in a given context, then 'Some A^M' can also be read informally as 'Some part of A', which can then be interpreted as follows:

(Sm/Mass) $\quad (\exists xA^M)\varphi x \leftrightarrow (\exists x/x \subseteq A)\varphi x.$

Also, because 'All A^M' is the dual of 'Some A', then 'All A^M', which, when A^M is a mass name, can also be read informally as 'Every part of A', is similarly interpreted as:

(All/Mass) $\quad (\forall xA^M)\varphi x \leftrightarrow (\forall x/x \subseteq A)\varphi x.$

Thus a sentence such as 'Water is a fluid', by which of course we mean 'All water is a fluid' can be symbolized as follows, where the predicate 'is a fluid' is based on the copula:

$(\forall xWater^M)Is(x, [\exists yFluid^M]),$

which by (All/Mass) and (Sm/Mass) and the identity theory of the copula reduces to:

$(\forall x/x \subseteq Water)(\exists y/y \subseteq Fluid)(x = y),$

which says that every part of water is part of some fluid.

Predication in the sentence 'Water is in the radiator' is not based on the copula but rather on the relation of being in the radiator, and of course it is not all water that is in the radiator, but only some water:

$(\exists xWater^M)In(x, [\exists_1 yRadiator]),$

which by the meaning postulate (Sm/Mass) and the extensionality (in the presumed context[55]) of the relation of being in the radiator reduces to

$(\exists x/x \subseteq Water)(\exists_1 yRadiator)In(x, y).$

---

55 There are other uses of the relation "in" that are not extensional, such as the relation of "in" between a story or someone's belief space and the propositions that make up the content of that story or that person's belief space.

What these last two examples also show is that predication is not always interpreted in the same way when mass nouns are the grammatical subject; that is, it may be universal as in 'Water is a fluid' or particular as in 'Water is in the radiator'.

Another type of problem arises when more than one determiner or quantifier phrase occurs in a sentence. The sentence 'There is some water in the radiator' has two determiners, for example, namely 'There is' and 'some'. Most logic texts would interpret this as above, i.e., as 'Some water is in the radiator', which amounts to ignoring the first quantifier phrase 'There is'. An alternative reading is possible, however, and is similar to our earlier example with the plural sentence 'There are twelve Apostles', which we interpreted as 'Twelve Apostles there are'. The sentence 'There is some water in the radiator' would be interpreted then as 'Some water in the radiator there is', which would be symbolized as:

$(\exists x Water^M/In(x, [\exists_1 y Radiator])[\lambda x(\exists z)(x = z)]^M(x),$

which by the meaning postulate (Sm/Mass) and λ-conversion reduces to the above analysis for 'Some water is in the radiator', which of course is what it should reduce to.[56]

What about when a mass noun is not the subject but the predicate of a sentence? In predicate position, the phrase 'is water' is interpreted as 'x is some water', which, semantically, is interpreted as 'x is identical with some water', so that the 'is' in this case is really the copula. Thus, e.g., the sentence 'The puddle is water' is analyzed as:

$(\exists_1 y Puddle)Is(y, [\exists x Water^M]),$

which, by the meaning postulates for Is and (Sm/Mass), reduces to:

$(\exists_1 y Puddle)(\exists x/x \subseteq Water)(y = x).$

Finally, we note that because the logic of classes as many contains an atomistic mereology, we have as a theorem of the system that combining two parts of water results in another part of water, i.e.,

$(\forall x/x \subseteq Water)(\forall y/y \subseteq Water)[(x \cup y) \subseteq Water]$

---

56 By the extensionality of $[\lambda x(\exists z)(x = z)]^M$, which is represented as follows:

$(\forall x)([\lambda x(\exists z)(x = z)](x) \leftrightarrow (\forall w/w \subseteq x)(\exists z)(w = z)),$

it follows that in the above context the predicate $[\lambda x(\exists z)(x = z)]^M(x)$ is vacuous and trivially equivalent to $x = x$. This predicate is provably identical with $[\lambda x Is(x, [\exists z])]^M$, in which the quantifier has been deactivated.

is provable in our logic, which of course parallels the same result for plurals. This is another respect in which mass nouns are like plurals. Another observation is that because being *part of* in our logic means being a subclass (as many) of, then it follows that mass nouns are provably distributive:

$$(\forall x A^M)(\forall y/y \subseteq x)(\exists z A^M)(y = z).$$

In other words, if x is some part of what a (nominalized) mass name A denotes, then every part of x is also some part of what A denotes. But let us note that even though a hydrogen atom and an oxygen molecule are in some appropriate sense "part of" some molecule of water, it does not follow that they also part of some water in our sense of *part-of*, namely $\subseteq$, which is based on membership in a class as many.

9. Concluding Remarks

Our basic thesis is that the unity of an assertion as a speech act, and of a judgment as a mental act, whether overtly expressed or not, consists in the mutual joint-exercise of an unsaturated referential and predicable concept in which each saturates the other in a kind of mental chemistry that results in an event, namely a mental act and a speech act as well if the mental act is overtly expressed. In our theory, referential and predicable concepts are unsaturated, complementary, intersubjective cognitive capacities that underlie our rule following abilities in the use of language, and in particular of our use of referential and predicable expressions. Referential expressions are represented in our logic by quantifier phrases restricted by a name, proper or common, and simple or complex. We have shown how our theory of predication applies to plural and mass noun reference and predication no less so than to predication in the indicative mode.

Our underlying logic for both plurals and mass nouns is a logic of classes as many as pluralities. The important new move in this logic is allowing a transformation of the names that are part of quantifier phrases into names that can occur as "terms", i.e., as arguments of predicates. What these new terms denote are pluralities, including the pluralities that consist of the minimal parts of what mass nouns denote. This logic has been developed in detail elsewhere, along with a set-theoretic semantics.[57] Both pluralities and the denotata of mass nouns (as, e.g., the various parts of water), are represented as being on the same logical level as individuals or single objects (each of which is itself a plurality or class as many of one), which is the way they are represented in natural language. This is

---

57 See Cocchiarella (2002).

different from most other treatments today of plurals by philosophers of logic. But as I have shown elsewhere, the principal alternative approach to the logic of plurals is reducible to the approach described here.[58]

References

Angelelli, I. (1979). "Class as One and Class as Many". *Historia Mathematica* 6: 305–309.

Boolos, G. (1984). "To Be Is To Be a Value of a Variable (or to Be Some Values of Some Variables)". *Journal of Philosophy* 81: 430–450.

Chierchia, G. (1998). "Plurality of Mass Nouns and the Notion of Semantic Parameter". In: S. Rothstein (ed.), *Events and Grammar*, Dordrecht: Kluwer Academic Publishers, 53–103.

Cocchiarella, N. B. (1986). *Logical Investigations of Predication Theory and the Problem of Universals*, vol. 2 of *Indices*. Naples: Bibliopolis Press.

Cocchiarella, N. B. (2001). "A Logical Reconstruction of Medieval Terminist Logic in Conceptual Realism". *Logical Analysis and History of Philosophy* 4: 35–72.

Cocchiarella, N. B. (2002). "On the Logic of Classes as Many". *Studia Logica* 70: 303–338.

Cocchiarella, N. B. (2007). *Formal Ontology and Conceptual Realism* (Synthese Library vol. 339), Dordrecht: Springer.

Cocchiarella, N. B. (2009). "Mass Nouns in a Logic of Classes as Many". *Journal of Philosophical Logic* 38(3): 343–361.

Cocchiarella, N. B. (2013). "Representing Intentional Objects in Conceptual Realism". *Humana Mente Journal of Philosophical Studies* 25: 159–174.

Cocchiarella, N. B. (2015). "Two Views of Plural Logic". *Studia Logica* 103(4): 757–780.

Eberle, R. A. (1970). *Nominalistic Systems* (Synthese Library), Dordrecht: D. Reidel Pub. Co.

Evans, G. (1982). *The Varieties of Reference*. Oxford: Clarendon Press.

Geach, P. (1980a). *Logic Matters*. Berkeley and Los Angeles: University of California Press.

Geach, P. (1980b). *Reference and Generality* (3rd edition). Ithaca: Cornell University Press.

---

58 See, e.g., Cocchiarella (2015). I want to thank Ignacio Angelelli and Woosuk Park for comments and corrections.

Kant, I. (1965). *Critique Of Pure Reason* (translated by N. Kemp Smith). New York: St. Martin's Press.

Laycock, H. (1972). "Some Questions of Ontology". *Philosophical Review* LXXXI (1): 3–42.

Montague, R. (1970). "The Proper Treatment of Quantification in Ordinary English". In: *Formal Philosophy, Selected Papers of Richard Montague*, edited and with an introduction by R.H. Thomason, New Haven: Yale University Press (1974), 247–270.

McKay, T. (2006). *Plural Predication*. Oxford: Oxford University Press.

Nicolas, D. (2008). "Mass Nouns and Plural Logic". *Linguistics and Philosophy* 31(2): 211- 244.

Normore, C. (1985). "Buridan's Ontology". In: J. Bogen and J. E. McGuire (eds.), *How Things Are*, Dordrecht: D. Reidel Pub. Co.,189–203.

Normore, C. (1990). "Ockham on Mental". In: J. C. Smith (ed.), *Historical Foundations of Cognitive Science*, Dordrecht: Kluwer Academic Pub., 189–203.

Pelletier, F. J. (1975). "Non-Singular Reference: Some Preliminaries". *Philosophia* 5(4): 451–465. (Reprinted in: F. J. Pelletier, *Mass Terms: Some Philosophical Problems*, Dordrecht: Reidel, 1979, 1–14).

Prior, A. N. (1967). *Past, Present and Future*. London: Oxford University Press.

Russell, B. (1903). *Principles of Mathematics*. New York: W.W. Norton & Co., Inc.

Scott, T. K. (1966a). *John Buridan: Sophisms on Meaning and Truth*. New York: Appleton Century-Crofts.

Scott, T. K. (1966b). "Geach on Supposition Theory". *Mind* 75: 586–588.

Spade, P. V. (2007). *Thoughts, Words and Things: An Introduction to Late Medieval Logic and Semantic Theory*, Version 1.2, http//hdl.handle.net/2022/18939.

Trentman, J. (1970). "Ockham on Mental". *Mind* 79: 586–590.

Ignacio Angelelli
The University of Texas at Austin

The Impact of Traditional Predication Theory on the Notion of Class[1]

Abstract: A text from the 18th c. Johann Caspar Sulzer is quoted where not only the notion of class (*classis*) is used but, moreover, a hierarchy of **orders** for classes is introduced. Such a hierarchy is developed not along the relation of membership but along the relation of inclusion. Given a class of order n, a class of order $n+1$ is not one which has the class of order n as a member but one of which the class of order n is a subclass. The main thesis of this paper (which is a modified, English version of my earlier *La Jerarquía*) is that the ordering of classes by inclusion rather than by membership is a consequence of the traditional, as opposed to the Fregean, predication theory.

Keywords: predication, class, orders of classes, Johann Caspar Sulzer

1. Classes and Predication

The notion of class depends on the notion of predication: a class consists of the entities of which a certain predicate is truly said. It so happens that in the history of logic two predication theories have successively prevailed: the Aristotelian, traditional one and then the Fregean, modern one.[2] It is easy to anticipate that these different theories of predication will generate different understandings of the notion of class.

Predicates have been generally regarded as composed of other predicates (e.g. *homo*, human, as consisting of *animal* and *rationalis*). The latter were often called *notae*, *Merkmalen*, marks of the concepts of which they are components. According to the Aristotelian paradigm predicates can be predicated of their marks ("some things are themselves predicated of others, but nothing prior is predicated of them; and some are predicated of others, and yet others of them, e.g. human of Callias and animal of human", *Analytica Priora* A 27), which is precisely forbidden by the Fregean model (marks of a concept are not predicated of the concept, e.g. animal is not said of human, contrary to the just quoted Aristotelian passage,

---

[1] I would like to dedicate this paper to the memory of my colleague and friend Angel d'Ors (1951–2012).
[2] Cf. my *Predication Theory: Classical and Modern*, as well as my *Studies*.

Grundlagen der Arithmetik, §53). In the modern model, one may predicate of the predicate human, for example, the predicate "having 7 billion individuals", i.e. predicates likely to be classified in traditional logic as "second intentions".

Thus, if the class of animals comprehends all those entities of which "animal" is truly said, it will obviously include, in the pre-Fregean framework, not only the individual animals, but **also** the subclasses of animal, e.g. the class of humans, the class of horses, etc., since "animal" is said not only of the individuals but also of the inferior predicates (universals, concepts) "human", "horse", etc. Therefore the traditional "class of animals" rather resembles the power set of the set of animals (upon a benevolent, previous replacement of the individuals by their unit-classes). All this can be anticipated from the famous Port Royal explanation of what is the extension of a concept:

> J'appelle *étendue* de l'idée les sujets à qui cette idée convient, ce qu'on appelle aussi les inférieurs d'un terme générale qui à leur égard est appellé supérieur, comme l'idée du triangle en général s'étend à toutes les diverses espèces de triangles[3] (I, ch. 6).

The class of triangles consists both of the individual triangles, written on paper or on blackboards, and of the class of equilaterals, the class of isosceles, etc. The Kneales, in *The Development of Logic* (1962: 318–319), comment on the Port Royal extension as if it was a mere confusion owed to the "metaphorical and unclear" use of the term "inferior", failing to observe that it is the entire conception of traditional predication what leads to include in the extension of a predicate the subordinate concepts along with the individuals (cf. my *Studies*, section 4.52).

The peculiarities of traditional predication theory should play a role as soon as the iteration of the relation of predication and the notion of a "predicate of a predicate" come into consideration. For one thing, in the classical understanding, predication lends itself to being regarded as transitive[4], which is ruled out in the modern approach. Both in the Aristotelian and in the Fregean approach one may arrange predicates into hierarchies, conceptual or even graphical as in the *arbor porphyriana*. While in the Porphyrian tree the highest predicates continue to be truly said of the individuals at the bottom, this is not the case in the modern hierarchy.

---

3   I call the extension of an idea the subjects to which the idea applies, also referred to as the inferiors of a general term, the latter being called, in relation to them, superior; as the idea of triangle in general extends to all the different sorts of triangles.

4   The traditional transitivity of predication made it difficult to accommodate second intentions in the *arbor porphyriana* – an "eleventh category" was occasionally created for that purpose, cf. Gasconius and my "Predication Theory: Classical and Modern".

2. Sulzer on Orders of Classes

The following text exemplifies the modern logic understanding of a hierarchy of "orders" of classes as constructed along the membership relation:

> Apart from separate individual objects, which we shall also call INDIVIDUALS for short, logic is concerned with CLASSES of objects; in everyday life as well as in mathematics, classes are more often referred to as SETS. Arithmetic, for instance, frequently deals with sets of numbers, and in geometry our interest lies not so much in single points as in sets of points (that is, in geometrical configurations). Now, classes of individuals are called CLASSES OF THE FIRST ORDER. Relatively more rarely in our investigations we come upon CLASSES OF THE SECOND ORDER, that is, upon classes which consist, not of individuals, but of classes of the first order. Sometimes even CLASSES OF THE THIRD, FOURTH, and higher ORDERS have to be dealt with. (Tarski 1994: 63).

In the history of traditional logic the notion of class is not common, let alone the division of classes into a hierarchy of orders. A very rare example is found in Johann Caspar Sulzer, making it clear, however, that Sulzer's orders are conceived along the relation of inclusion among classes, not of membership. The original Latin is quoted first, followed by an English version:

2.1 Latin text

Cap. 1 De conceptibus et quomodo in species et genera ordinentur.

§1 Quicquid in rerum natura sensibus nostris observatur, singulare est.
§2 Omne singulare ab alio singulare differt eo ipso, quod est aliud, et individuum vocatur.
§3 Omnia itaque in rerum natura sunt diversa, et licet sint innumerabilia, tamen semper unum non est alterum: unum individuum non est alterum individuum: Jacobus non est Johannes.
§4 Quamquam vero diversa sunt, in multis tamen convenire deprehendimus. In quibus convenient vel similia sunt, eo respectu ad eandem referuntur classem: in quibus non conveniunt, ad diversam classem referuntur.
§5 Quae ad eandem classem pertinent, idem nomen habent: diversa classis diversum nomen obtinet, et nomen unius classis alteri non convenit. Vgr. Jacobus et Henricus, quia similes sunt, pertinent ad unam eandemque classem, cui virorum nomen est. Hinc uterque est vir. Anna vero et Maria, quia multo differunt a Jacobo et Henrico, diversam classem constituunt, quae nomine feminarum gaudet. Est ergo et Anna et Maria femina, sed vir non est femina nec femina vir.

§6 Ex collatione igitur rerum omnium plurimae classes rerum oriuntur, et semper diversae, quia classis una non est altera, §2, §3.

§7 Classes, quae ex similitudine individuorum promanant, erunt classes primi ordinis, et dicuntur species, vgr. mares, feminae.

§8 Classes primi ordinis, licet diversae, similitudinem quondam habent, quae similitudo exhibet classes secundi ordinis, seu classes specierum, et tales classes vocantur genera. Sic mares et feminae ex Jacobo, Henrico, Anna, Maria orti homines nominantur, hinc mares et feminae erunt species, homines vero genus.

§9 Classes, quae ex similitudine classium secundi ordinis oriuntur, tertii ordinis classes producunt, et generum superiorum nomine veniunt. Tale genus superius est, si homines et bruta conferuntur, et ex eo, quod utrinque vivunt et sentiunt, in unam classem ordinantur, quae classis animalium nomen habet, animal ergo erit genus superius.

§10 Ulterius sic procedendo classes quarti, quinti, cet. ordinis exhibentur, et sic superiora semper genera nobis sistuntur, donec tandem ad unam ultimam classem perveniamus, quae supremum genus dicitur, quod ens est […]

§13 Quod itaque genus nomen habet, et quidquid de genere dicitur, id nomen quoque habent omnia, quae sub hoc genere comprehenduntur ad individua usque, et de iis quoque absolute dicitur. Exemplum sit animal, quia homines, bruta, mares, feminae, cet. comprehendit, dicetur homo est animal, brutum, vir, femina, Jacobus, Maria, cet. est animal. Et si dicitur: animal sentit, etiam homo, brutum, vir, femina, Jacobus, Maria, sentit.

2.2 English translation

Chapter 1. Concerning concepts and in what way they are ordered in species and genera.

1. Whatever we observe by means of our senses in reality is singular.
2. Every singular differs from another singular, by the very fact that it is another, and is called an individual.
3. Thus all things in reality are diverse, and although they are innumerable, nevertheless, one is never another one individual is not another individual: Jacobus is not Johannes.
4. But although they are diverse, in many respects however we discover that they agree. In virtue of that respect in which they agree or are similar, they are referred to the same class; in virtue of that respect in which they do not agree, they are referred to a distinct class.

Impact of Traditional Predication Theory on the Notion of Class 97

5. Those things which pertain to the same class, have the same name; a different class obtains a different name and the name of one class is not suitable for another. For example, Jacobus and Henricus, since they are similar, belong to one and the same class, of which the name is "man". Hence each is man. But Anna and Maria, since they differ greatly from Jacobus and Henricus, constitute a diverse class, which bears the name woman. Therefore both Anna is woman and Maria is woman, but man is not woman nor woman man.
6. Thus from the collation of all things many classes of things arise, and they are always diverse, since one class is not another.
7. Classes which are formed from the similitude of individuals, will be classes of the first order, and are called species, for example: males, females.
8. Classes of the first order, although diverse, have a certain similitude, which similitude yields classes of the second order, or classes of species, and such classes are called genera. Thus males and females, derived from Jacobus, Henricus, Anna, and Maria, are named humans, hence males and females are species, but humans is a genus.
9. Classes, which are derived from the similitude of second order classes produce classes of the third order, and go by the name of higher genera. Such a superior genus we have when humans and beasts are compared and from the fact that both live and sense, they are ordered in one class, which class has the name "animal"; therefore animal will be a higher genus.
10. Proceeding further, thus, classes of the fourth, fifth, and the other orders are reached, and thus even higher genera are established for us, until at last we may arrive at the one ultimate class, which is called the highest genus, namely being […].
13. Whatever name then the genus has, and whatever is said of the genus, that name have also all the things that are comprehended under the genus, all the way down to the individuals, and is said of these absolutely as well. Let animal be an example; because animal comprehends humans, beasts, males, females, etc., it will be said "human is animal", "beast, male, female, Jacobus, Maria, etc. is animal". And if it is said, "animal senses", then also human, beast, male, female, Jacobus, Maria, senses.

3. Comments

Sulzer starts from classes of individuals: the class of men: {Jacobus, Johannes…} and the class of women: {Anna, Maria…}. Such classes of individuals are **first order**. By comparing first order classes, and noticing their similitudes (the class of men resembles in certain respects the class of women but not, for example, the class of trees), new classes, now referred to as **second order**, are formed. Contrary to the expectations of a modern reader, however, the new second order class does not contain, as elements, the class of men and the class of women. The new second order class is just a class more inclusive than the class of men and more inclusive than the class of women. The second order class of humans, "comprehends" (such seems to be Sulzer's favorite term to designate the relationship between the class and whatever we want to dump into it) all of its subclasses down to the individuals (cf. §13). If the individuals are replaced by their unit-classes, then **comprehend** is exactly the converse of class-inclusion, and generally the order of a Sulzerian class is higher than the order of another class iff all the classes included in the latter are included in the former.

4. Predication: Central Notion

That the notion of predication has been subject to two versions – traditional and modern – with important differences (e.g. in one version predication is transitive but not in the other version, one version leads to construe orders of classes as in Sulzer whereas the other version as in Tarski, etc.) should not make us lose sight of the fact that what is common to the two versions – simply, predication – has been and continues to be the central topic of logic and ontology (or metaphysics) in the entire history of philosophy, from Antiquity till now. Now, what is common to the two versions is exactly the Fregean notion of predication, which Frege obtained by removing a number of spurious components from the traditional notion (subordination, identity…), just keeping, as predication, the converse of the sheer falling of an object under a concept.

References

Angelelli, I.[5] (1965). "Leibniz's Misunderstanding of Nizolius' Notion of *Multitudo*". *Notre Dame Journal of Formal Logic* VI: 319–322.

Angelelli, I. (1967). *Studies on Gottlob Frege and Traditional Philosophy*. Dordrecht: Reidel.

Angelelli, I. (1974). "La jerarquía de clases de Johann Caspar Sulzer (1755)". *Cuadernos de Filosofía*, Universidad de Buenos Aires, XIV, 21: 90–94.

---

5   The Angelelli items listed here are available online (for *Studies* only selected sections) at: http://www.utexas.edu/cola/depts/philosophy/faculty/iaa4774.

Angelelli, I. (1975). "Freges Ort in der Begriffsgeschichte". In: Ch. Thiel (ed.), *Frege und die moderne Grundlagenforschung*, Meisenheim am Glan: Verlag Anton Hain, 9–22.

Angelelli, I. (2004). "Predication Theory: Classical and Modern". In: H. Hochberg and K. Mulligan (eds.), *Relations and Predicates*, Frankfurt: Ontos Verlag, 55–80.

Aristotle. (1949). *Aristotle's Prior and Posterior Analytics*. A revised text with introduction and commentary by W. D. Ross, Oxford: Oxford University Press.

Frege, G. (1884). *Die Grundlagen der Arithmetik*. Breslau.

Gasconius: Ioannii [sic] G*asconii bilbilitani, liberalium artium magistri, atque in Oscensi academia publici professoris, in logicam, sive dialecticam Aristotelis Commentaria*, Oscae, Excudebat Ioannes Perez a Valdivielso, Typographus Universitatis, anno 1576.

Kneale W. and Kneale M. (1962). *The Development of Logic*. Oxford: Oxford University Press.

Port-Royal. *La Logique ou l'Art de Penser*. Cinquième édition, chez G. Desprez, Paris 1683.

Sgarbi, M. (2012). "Immanuel Kant: Die falsche Spitzfindkgeit der vier syllogistischen Figuren". In: S.-K. Lee, R. Pozzo, M. Sgarbi and D. von Wille (eds.), *Philosophical Academic Programs of the German Enlightenment: A Literary Genre Recontextualized*, Stuttgart-Bad Cannstatt: Frommann-Holzboog, 189–210.

Sulzer, J. C. (1755).[6] *Facies nova doctrinae syllogisticae, qua multo plures modi figurarum syllogisticarum facillimis et certissimis regulis proponuntur quam hactenus exhibiti sunt, ex qua omnes syllogismos, cujuscunque sint conditionis, sine ulla inmutatione conclusionis in quavis figura exhiberi posse, demonstratur*, Tiguri, ex officina Heideggeri et Soc.

Tarski, A. [1941] (1994). *An Introduction to Logic and the Methodology of the Deductive Sciences*. New York and Oxford: Oxford University Press.

---

6   As noted in the preface, Sulzer was a professor at Winterthur: *rector scholarum vitoduranarum*. The preface was signed in that Swiss city, in *Kalendis majis* 1754. Quoted text is from a film (at The University of Texas-Austin library) of an original at the Staatsbibliothek Bamberg. A copy is also found at the Bayerische Staatsbibliothek, and available online: http://www.mdz-nbn-resolving.de/urn/resolver.pl?urn=urn:nbn:de:bvb:12-bsb10044529-8. Sulzer had at least one distinguished reader; according to Sgarbi, *Facies nova doctrinae syllogisticae* "is the only work of Aristotelian logic found in Kant's library".

Allan Bäck
Kutztown University, Pennsylvania

The Future of Predication Theory

Abstract: In the history of predication, there has come to be an emphasis on impersonal statements, a move from speaker meaning to sentence meaning, an elimination of intentionality and standpoint via the elimination of indexicals, and a move from ordinary to ideal language. Following these trends in light of their historical success, I suggest that we should develop a theory of predication with the following features: formalism, having a sharp distinction between the pure theory and its interpretations, and the primacy of two-place relations. Accordingly, I offer a theory of predication with the following features: an uninterpreted language, based on the combinatorials of primitive symbols and on the application of formation rules so as to give well-formed formulae (wffs), construction of *n*-place predicates via unsaturating the wffs, the reduction of all n-place predicates to two-place predicates (relations). I end with considering how successfully such a predication theory may be interpreted and applied, in particular to metaphysics.

Keywords: predication, combinatorial, dyadic relation, formalism

0. Introduction

From a logical point of view, the theory of predication has moved over time from one reflecting the surface grammar of certain natural languages, predominantly but not completely Indo-European, to one more in tune with the deep structure of relations and indeed of the world, at least as our current scientific theory presents it. In Aristotelian terms, we have moved from what is most evident to us but least evident in itself to one that is (largely) most evident in itself but least evident to us.

Continuing these trends, I suggest that we should develop a theory of predication with the following features:

- Formalism
- A sharp distinction between the pure theory and its interpretations
- The primacy of two-place relations

Accordingly, I shall propose a theory of predication with the following features:

- An uninterpreted language, based on the combinatorials of primitive symbols and on the application of formation rules so as to give well formed formulae (wffs).

- Construction of n-place predicates via unsaturating the wffs
- The reduction of all n-place predicates to two-place predicates (relations)

In short, I shall present a purely formal theory where predication has the basic syntactic structure of a two-place relation. Various configurations of that relation constitute other n-place predicates, both monadic and polyadic.

In order to motivate an interest in my theory, I shall begin with some historical remarks.

1. Historical Perspectives

I do not propose to give a history of predication and its theory, even for the Indo-European languages. Yet, in order to motivate the theory that is to follow, let me just note certain historical trends in the language of science and philosophy and its theory, without offering much support.

- An emphasis on third-person – or better: impersonal – statements, and a de-emphasis on statements of others, like first- and second- and dual.

In the *Theaetetus* Plato sticks to simple statements like 'Theaetetus flies' [263C]. In *On Interpretation*, a seminal work for both grammatical and logical theory, although Aristotle recognizes other grammatical types of statements, he takes a statement to affirm or deny one thing of another [17a25; 17a1]. Aristotle uses this type of declarative statement in his syllogistic and theory of demonstration, although in his *Topics* he does extend the doctrine to dealing with interrogatives [*Topics* VIII.1]. The declarative sentence then becomes the standard type discussed in logical theories of predication.

Medieval times still had some literary forms involving dialogue and conversation: above all, the Latin medieval with their *disputationes* and *obligationes*. Yet, although appeals to the authority of the Fathers were made in the *Sentences* commentaries as well, the persons seem to drop out. Rather, the content of the propositions mattered. Although Plato had protested in "The Seventh Letter" that written philosophy is dead philosophy, the written text, with its permanence and relative immunity to context starts to prevail over oral discussion.

With the printing press, reliance on the written treatise and an emphasis on content over speaker become the norm. It no longer mattered whose theory it is; what mattered is the theory. Even Descartes, when he speaks from the first-person perspective in the *Mediations*, does so from a quite impersonal point of view: anybody can take up that point of view. Today, with reliance on recordings, holograms, and avatars, the content of the speech act has become even more divorced from a particular time and place and speaker.

- A move from ordinary to ideal language

Already with Plato, we have a distaste for and a deviation from the common ways of speaking and thinking of the *hoi polloi*. In developing his own position, Aristotle develops a technical vocabulary that departs from common usage.[1] In this sense, at least, Aristotle's thought is developmental: starting from ordinary language, he is creating his technical language so as to make clear what is really going on. He proceeds by eliminating respectable, endoxic ways of talking that do not match up with reality and then by introducing new, contrived ways of talking to fill in the gaps: methods of subtraction and addition. This type of constructing an ideal language is semantic, in the sense that Aristotle wants to admit only terms that refer directly (or, if indirectly, that can be parsed away) to objects *in re*.

For instance, in discussing paronymy in *Categories* 1, Aristotle makes the abstract term, like 'bravery', basic and the concrete one, like 'brave', derivative, even though this flouts the surface grammatical structure. He again complains that ordinary language misleads in not having suitable derivative terms, as with the concrete paronym, taken as the *quale* for the quality named by 'virtue' (ἀπὸ τῆς ἀρετῆς ὁ σπουδαῖος).[2]

Aristotle makes reforms also of a syntactic type, by subtracting some grammatically respectable construction and then adding on some new ones. For instance, he legislates away unnatural predication and introduces the strange 'ὑπάρχει' ('belongs') structure of predication. Aristotle wants his logical subjects to be substances and not accidents. He says that individual subjects should not be predicated, even while admitting that we commonly do this:

> It is clear that some beings are naturally said of no beings. For nearly every perceptible is such as not to be predicated of anything except *per accidens* (κατὰ συμβεβηκός). For we say that the white [thing] is Socrates, and the one who is approaching is Callias.[3]

Aristotle admits that in ordinary discourse singular (terms) are predicated, but must be then "predicated *per accidens*".[4] He gives examples like 'the white (thing) is Socrates', where a name signifying a primary substance is predicated of a

---

1 See Bäck (2000: 144); Bäck (forthcoming).
2 *Cat*.10b7.
3 *An. Pr.* 43a32–6, my translation; cf. *Soph. El.* 179a39-b2; Ammonius (1897: 53,22–8).
4 Likewise Simplicius (1907: 51,13–8) admits that 'Socrates is Socrates' is a true predication, but denies it to be an instance of the 'said of' type. Instead, he analyzes it as an

paronymous expression derived from the name of a quality, 'whiteness'.[5] But this way of talking does not match reality:

> One must look for truth from the realities. It holds thus: since there is something itself that is predicated of another not *per accidens* – I mean by *per accidens* as when we sometimes say that that white (thing) is a man, this not being similar to our saying that the man is white: for he is not white as being something different, while the white (thing) [is a man] because being a man has happened to white – thus some are such as to be predicated *per se*.[6]

In medieval times the move from an ordinary to a universal, ideal languge continued. Those like al-Fārābī accordingly saw quite different roles for logic and grammar:

> Grammar shares with it [logic] to some extent and differs from it also, because grammar gives rules only for the expressions which are peculiar to a particular nation and to the people who use the language, whereas logic gives rules for the expressions which are common to all languages.[7]

Thus he advocates deriving the paronyms systematically, by using a modified version of the Arabic grammatical form of the *maṣdar* (the verbal infinitive).

Likewise Avicenna notes the varieties of the forms of the *maṣdar* in Arabic too. He says that sometimes the *maṣdar* does not have a special expression but instead just uses an absolute name, as with 'motion' (*taḥarruk*). So sometimes the *maṣdar* is the basic form grammatically; sometimes it is not.[8] The logician may have to modify, or mangle, ordinary language.

Likewise those in the medieval West like Anselm worked in this tradition. Anselm comes up with the startling pronouncement, that '*grammaticus est grammatica*' ('the grammatical is grammar') is true strictly speaking.[9] As D. P. Henry puts it, '*grammaticus*' signifies precisely *grammatica* but appellates or names man.[10] That is, '*grammaticus*' signifies the quality grammar. It appellates or stands for things

---

    instance of the name's being predicated synonymously of Socrates, as opposed to the predication of "one thing of another".

5 'The white' sometimes means 'whiteness', but it cannot do so here, for 'whiteness is Socrates' is false.

6 *An. Po.* 81b22–9, my translation, which differs significantly from Barnes' here.

7 Al-Fārābī (Dunlop (1956: 228,8–10 [trans, p. 233])); "Al-Fārābī's Paraphrase of the *Categories* of Aristotle," (Dunlop (1958: 172,28–173,8 [trans. p. 187 §9])): Wright (1967, §358).

8 Avicenna (1970: 26,9–20).

9 Henry (1964: 4.233; 4.6,10).

10 Henry (1964: 119).

presently existing *in re*, namely, for the quality grammar, with the added feature of being in a presently existing subject. Still it just means grammar. A grammatical person is just one of the places where grammar is present. Anselm is helped to this conclusion by the common, endoxic habit, even in Aristotle, of referring to items in accidental categories by derivative terms like 'the grammatical'.[11] To hold that *grammaticus* just is grammar violates the common way of speaking. Anselm recognizes this but doesn't mind.

Still Anselm seems not to want to subscribe fully to constructing an ideal language, "[r]*ather, Anselm wants to analyze and order the ambiguities of natural language while retaining it and its nuances.*"[12] As Marilyn Adams says, "*it is not part of Anselm's goal to reform usage: that is already given, fixed centuries ago in the texts themselves.*"[13] Instead, Anselm seeks to clarify improper usage and develop skills for construing texts.

Later on, Ockham explicitly embraced a program of constructing a protocol language. He seems to seek to construct a conventional language directly reflecting the structure of the mental language, which in turn reflects, and indeed comes from, the real order of the world.[14] Thus he contrasts what holds *de virtute sermonis* and *secundum proprietatem sermonis* from ordinary usage, even that in Scripture. In the latter way, Ockham says, *Timothy* I.6 may say that God has justice, but strictly speaking that claim is false; rather, as Anselm says in *Monologion* 16, God is justice.[15] Ockham has the mental language containing both abstract and concrete terms in the category of quality. The abstract terms, like 'whiteness', signify the qualities themselves. The concrete terms, like 'white', signify individual substances primarily and qualities secondarily, since, as Aristotle says, a paronym like the white is a complex, the substance having whiteness.[16]

In later times, as the Arabic algebra came to be applied more and more, and as the scientific vocabulary develops, the language became more and more removed from how people speak ordinarily. New notation was introduced, including the identity sign (R. Recorde) and the integral sign (Leibniz); new words came about, like 'phlogiston', 'oxygen', and 'galaxy'; old words had their meanings transformed,

---

11 *Cat.* 1b29.
12 Sweeny (2012: 97).
13 Adams (2000: 109)
14 Panaccio (1991: 98) discusses how for Ockham detailing the mental language amounted to constructing an ideal language.
15 Ockham (1974), *Summa Logicae,* I.7.157–75.
16 Ockham (1970), *Ordinatio* I, Prologue q. 1; Panaccio (2004: 64); Gaskin (2001: 256–261). So too Burleigh: cf. Klima, (1991: 78–106).

like 'energy' and 'planet', 'quark' and 'strange'. These days, those in one subfield of physics can hardly understand the writings from another subfield.

- A move from speaker meaning to sentence meaning – in Latin medieval terms, a focus on the proposition rather than on the *usus loquendi*. Accordingly, the social context of language drops out – so far as is possible.

In the *Categories* Aristotle worries that statements, like substances, persist through time while receiving contraries. Thus 'Theaetetus sits' can be at one time true and at another false. Aristotle says that, when substances receive contraries, the substances themselves change; when statements receive contraries, the statements themselves do not change, but rather the objects, the things that are the truth-makers (or falsity-makers) for those statements, change [*Cat*. 4a20-b4]. Aristotle views a statement as a single thing persisting through time.

Why does Aristotle then think it at least plausible that the uttered statement remains the same while taking on contrary truth values? The speech act does not remain the same speech act. It seems that what remains the same is only the content of the utterances: what those speech acts all assert. All those utterance make the same statement, that Theaetetus is sitting. While that statement remains the same, the objects that it signifies, the *pragmata*, the facts of the matter or state of affairs if you will, may change, as Aristotle says. In terms of its content, the statement concerns an individual human substance and his position at the time of the utterance.

Then what remains the same in all these utterances of 'Theaetetus is sitting' is what they all assert: the content of the statement – more or less what Frege later would call its sense, and the medievals signification. This content is what the statement is. In Aristotle's mental language that content would be a quality. In the spoken language, the content would be a certain order of significative sounds and a quantity; likewise for statements in the written language with reference to inscribed marks. Aristotle then starts with statements in use, but ends up dealing with them apart from that use.

Later logicians continued to focus on the content of the utterance, abstracted from its context. With al-Fārābī and Averroes and especially with Avicenna, the dialectical strand of live conversation loses its philosophical luster; at best, it offers preliminary training for the student of philosophy. An active *noûs* suffices instead. Dialectic then becomes the method for dealing with ordinary, non-philosophical discourse. Understanding the intention of the speaker and the customs of the speakers, the *usus loquendi*, here becomes crucial in interpreting such discourse. However, philosophical insight and wisdom come much better from the technical discourse of the philosopher and the active intellect. Contemplating eternal truths, grounded ultimately in the divine intellect, left little need for the progression of discourse between speakers.[17]

- Elimination of intentionality and standpoint via the elimination of indexicals

In line with these trends, the stance of the speaker and author loses its importance. Statements begin to assume a timeless character, which they did not have before, even with Aristotle. The rising precision of calendars and clocks made it possible to state, timelessly, a proposition via including a time index within the statement itself. Leibniz himself may not have done so with the predicates in his monads (as for him time is merely phenomenal), but later versions of monads do so routinely.

When Descartes gave us a coordinate system, it came to be possible for formulate events as occurring in a space-time from a timeless point of view, as with Newton'a absolute space and time. Relativity physics, with its reference frames given without regard to the time at which the scientist states them, again eliminates the need for us to have indexicals like 'here' and 'then' in our scientific vocabulary. In science, we have a change from there being different now's at different times, as with Aristotle, to there being a single now moving along a timeline. McTaggart had argued that stating temporal relations timelessly (his B series) presupposes a temporal process of change (his A series). But now there seems to be no need for indexicals like *'two days earlier than now'*. Rather time and even change can be given up altogether, as some current physicists suggest.[18]

- The rise of a pure formalism

Offhand, abstracting from as much material content as possible seems mistaken. Although Aristotle bases his sciences on abstracting various features of individuals, so as to distinguish them *qua* movable, *qua* numbered, *qua* shaped etc., he did not want to abstract away from all material content. He found it to be a category mistake, to think of geometrical shapes in terms of their mathematical structure: he has a nice argument that analytical geometry is impossible [*An. Po.* I.7]. Too Heidegger found abstraction generally alarming: moving us away from engagement with Being (*Dasein*) to seeing objects in the world as this and as that – even though we may be doomed to do so: we cannot hold many items in active consciousness and must frame problems in simpler terms in order to discuss them or even to think about them. So abstraction has its dangers.

---

17 Bäck (2015).
18 Barbour (2000); Lockwood (2005).

Be that as it may in ordinary consciousness and life, in science abstraction has long been the way to go. Analytic geometry and coordinate systems made it possible to construe physical objects as sets of space-time points. Abstracting away from irregularities of textures and instead thinking of frictionless surfaces made great progress in projectile motion; construing orbiting objects as point masses aided the development of celestial mechanics. The current standard model for atomic particles and the Grand Unified Theories (GUTS) stand in this tradition. Again in anthropology focusing on abstract relations has shown the similarity of structure of seemingly disparate kinship customs.

Likewise in mathematics, abstraction from content has made possible the unification of many fields in terms of Galois groups and category theory. Here the idea is to keep the structure of relations and delete the content and natures of the things being related. So too in logic, focusing on the purely formal, uninterpreted system have the development of many logical systems, each of which may then be interpreted in applied in a number of ways. The case of propositional logic with its truth trees (analytic tableaux) has famously made digital computing possible, through its twin applications to hardware and software.

2. What We Want in a Theory of Predication

Given the past, what features should we like a theory of predication to have in the future?

- Formalism

Well, given the success of formalism over the last few centuries, I see no reason why it should not be pursued also in a theory of predication. Modern linguistics has long been moving in this direction by postulating various deep structures of abstract patterns having various natural languages as their instances.

- A sharp distinction between the pure theory and its interpretations

Even today, the development of pure theory or even pure scientific research does not look practical to many people. Yet, once again, the historical evidence looks overwhelming. It has just turned out that having a large store of pure theory, as divorced from content as possible, bears a lot of fruit both in theory and in practice. The theoretical advantage of an uninterpreted theory comes from being able to develop possibilities that do not look at all plausible from the viewpoint of our current intuitions. Yet those intuitions will change, along with theoretical advances.

Pure science also yields practical applications, although perhaps not immediately. Think of number theory and its focus on prime numbers. This had little application until recently, in encrypting data. Imaginary numbers were a curiosity until their use in describing and designing electrical circuits. Fuzzy logic now is used in reconstructing video images; 3-valued logic has come to have its uses in quantum theory and in computer chip design.

- The primacy of two-place relations

Formally, 2-place relations seem irreducible. For instance, consider truth-functional relations.[19] We can express all of them by a 2-place relation like the Sheffer stroke. Still these 2-place relations cannot be expressed by a 1-place relation like the negation operator.

Moreover, in reality, if we are to have fairly stable, complex objects, such as life forms are, we need relational, hierarchical structures. Complexity theory has suggested that dyadic relations make such stable structures possible. N-adic relations, n >2, produce structures with poor stability. (A merely monadic world would lead to no change and to a timeless monism.)[20]

So then, formally as well as materially, we seem stuck with some relations, the two-place ones. But do we need n-adic relations, n >2, or properties, the so-called 1-place relations also? No, as I sketch below.

3. A New Theory of Predication

Accordingly, I propose a theory of predication with the following features:

- A purely formal, uninterpreted language, based on the combinatorials of primitive symbols

Take some language L with some set of primitive symbols (whose complexes may then be abbreviated by other symbols. Using the Kleene star operator, generate all *possible* sequences of those symbols. In practice, we stop at finite or enumerable sequences – even at relatively small finite sequences, as those are all that we in fact can handle.

- Formation rules so as to give wffs in that formal language L

---

19 Logical systems having different numbers of truth values can still be expressed dyadically, via using conjunctive normal form.
20 Lewin (1992: 45).

A language is constituted in part by its formation rules. Yet often we have a choice about what formation rules to use. For example, a modal system may allow for iterated modalities by including, or not including, a formation rule like:

If Φ is a wff, then □Φ is a wff.

But other formation rules yield other results. Aristotle seems to presuppose a Necessity Introduction rule like:

If P belongs to every S in virtue of its essence (*per se*), then P belongs by necessity to every S

Here there is no place for iterated modalities (so long as we do not allow: P belongs by necessity to every S, then P belongs to every S *per se*).

So there are various possible sets of formation rules that may be applied to the sequences generated from a set of primitive symbols. Each such set will generate wffs in some language. Languages working from the same set of primitive symbols may be aid to constitute a *language family*.

Why bother making this point about language families? Logicians are human beings starting from a natural language. They then construct and apply various logical languages. Over time some features that were discarded or neglected may come to be included in the formal language. For instance, mathematicians did this with irrational numbers and then with negative square roots; logicians did this with iterated modalities. I want to leave it open – and emphasize – that various features in a formal language may be modified and included.

In regard to predication, this means that some predicates may seem to us to be ill-formed now, but later on may be admitted. For example, a predicate having a logical operator as its argument; a predicate having a grouping operator as its argument.

In a language, relative to its formation rules, there will be some sequences of symbols that are well-formed formulae (wffs); consider and eliminate all other sequences of symbols as ill-formed.

- Construction of n-place predicates via unsaturating the wffs

Remove one symbol, or all tokens of a type of symbol, and replace with a variable x_j. (first order) or Φ_j (second order) or… This operation may be repeated n times: the nth variable will be x_n. for instance, if we start with the fully saturated wff 'Raa', we get:

$Rx_1a\ Rax_1\ Rx_1x_1\ Rx_1x_2\ Rx_2x_1$
$\Phi aa\ \Phi x_1a\ \Phi ax_1\ \Phi x_1x_1\ \Phi x_1x_2\ \Phi x_2x_1$

(Unsaturating via removing variables and replacing them with other variables seems useless but harmless.)

In theory any such symbol may be removed, so as to constitute a predicate. In practice, we allow for only predicates stated in the formation rules so as to generate wffs. Thus in predicate logic there is the wff

(x)(Fx ⊃ Gx)

Formally, once we abstract away from the particular set of formation rules that are normally used, we may replace '⊃' or even '(' by a variable. The remaining complex will be a predicate; the variable may be replaced by a symbol (the same or different); saturation will lead to a proposition, which may or may not be well formed, relative to some formation rules.

Each variable needing saturation is a "place" in the "n-place predicate". In practice, we may just ignore some of these places and speak of the places only of a certain type or level. Hence we may speak of '$\Phi x_1 x_2$' as 'a 2-place predicate', even though it has 3 variables. However, we need not speak thus.

What advantage do we gain by allowing such "wild" predicates at first, especially when in practice we then move to customary practice? Well, first, we thereby make explicit what assumptions we are making in order to get the wffs and respectable predicates in the system that we end up with – and what possibilities we are ruling out. Second, more fundamentally, allowing for such wild predicates enables a deeper abstraction and a greater covering power of the theory. Replacing '⊃' by a variable enables us to talk about all 2-place logical connectives or functions in, say, predicate logic. Replacing ')' by a variable enables us to talk about various grouping relations generically.

- The reduction of all n-place predicates to two-place predicates (relations)

Make predication a dyadic relation, between two objects, as in 'Rxy'. In contrast, traditional predication is monadic, between one object and a predicate, as in 'Fx'.

Other relations can be eliminated, by restating or parsing in terms of the 2-place relations.

N-adic relations, $n > 2$, can be reduced to the dyadic in various ways.

(1) There is Quine's method of dividing a fully saturated relational complex into two parts, a constant or variable and the rest and speaking of the relation between those two parts.[21] For instance, take '$Rabc$' and construe it as '$Rabx$'.

---

21 Quine (1954: 180–2); (1966: 241, §41). Russell (1985: 35) rejects such reductions, but he does not have Kuratowski's method.

(2) *N*-adic relations, *n* >2, can be reduced to conjunctions of 2-place relations. To reduce *n*-place relations, *n*>2, to 2-place relations, consider their extensions. For instance, a 3-place relation has an extension consisting of a set of 3-place sequences: $\{<x_1, x_2, x_3>, <x_1, x_2, x_3> \ldots\}$ Each such sequence can be reduced to a nesting of 2-placed relations.

First, reduce a 2-place sequence to a nest of 2-place relations, via the Kuratowski method: define '<x, y>' as '{{x}, {x, y}}'. Now construe a n-place sequence as a union of two sequences: $<x_1, \ldots, x_{n-1}>$ and $<x_n>$. Use that method to reduce it. Continue to use the method on the former, $<x_1, \ldots, x_{n-1}>$, until the entire n-place sequence is defined in terms of 2-place relations.

So we do not need *n*-adic relations, *n* >2. We can get by with just 2-place relations. Yet these cannot be reduced to 1-place relations or properties.

On the other hand, those 1-place relations or properties can likewise be reduced to 2-place relations. Why bother to do so? For one thing, in theory construction, we prefer ontological parsimony: as few basic types of things as possible.

So I show in general how to reduce monadic properties to dyadic relations. I offer two methods: one (having two stages: first intensional and then extensional) theoretically satisfying but impossible for us to apply fully in practice; the other possible to apply but theoretically lacking.

A. (i) [intensional] Start with equivalence relations. As equivalence relations are transitive etc., we get sets of 2-tuples, each of which satisfies a given equivalence relation R_i. So $R_i = \{<x_1, y_1>, <x_2, y_2>, <x_1, y_2>\ldots\}$ The property P_i is the property of being in the set R, sc., of some thing belonging to that set.
An equivalence relation [of the sort discussed in A(1)] then gives a criterion for inclusion in one or more of those sets generated by the functions, R_i. Consider the set of all logically possible equivalence relations. For each set generated there will be one or more such equivalence relations. With luck, we might find one, one useful for our purposes.

(ii) [extensional] Let $P_{i'} = \{x_1, y_1, x_2, y_2\ldots\}$ be the set of items in R_i. Each such set is the extension of a property. To say that a thing has a property is to say that it belongs to such a set. (Note that 'x belongs to y' is another dyadic relation.) Bob Hale calls this method "grounding" a property on a relation.[22]
This can be reversed: take a property P and its extension, $P_{i'} = \{x_1, y_1, x_2, y_2\ldots\}$. Then take the set of all ordered pairs of elements of that set. That will yield

---

22 Hale (1987: 59).

$R_i = \{<x_1, y_1>, <x_2, y_2>, <x_1, y_2>, ...\}$ This works because R_i is based on an equivalence relation.

Consider a 2-place function, from the (universal) domain to the set of individuals satisfying some relation, which will be a subset of the domain, like the set of goats or the set of hunks of copper or the set of yellow things or the set of sets of sets of 2 things. This is a 2-place function or relation going from the individuals in the domain U to a range R. Now one such function will give just the set associated with a certain extension of a property. So we get:

$f_i(d_1, d_2, ...) \Rightarrow \{r_1, r_2, ...\}$, the set R_i, where r_i is a sequence or set of n members, $n \geq 0$,

This function requires allowing for a universal domain U containing all *possible* individuals and relations. (Some people don't want such "inflationary" domains.[23]) As such domains are routinely uncountable, we can only postulate but not use such a function, as with the Choice function in ZF set theory. The function itself names the property, be it one-place like yellowness or generosity, or be it two-place like being 2.[24]

Either method works formally. We can just stipulate that there always is an equivalence relation for each property or that our domain contains all possible individuals and their relations to be admitted. Indeed, once we admit the existence of the set of the extension of the property (even if we cannot know or list all of its possible or even actual members), we can define that relation. Again, once we have the universal domain U and that extensional set, we can get the function mapping the domain onto that range R_i.

B. In practice there is another option for defining an equivalence relation, one that is often more conceptually satisfying and heuristically fruitful, at first anyway. More generally, take some particular thing as the exemplar. Then use the equivalence relation. One might construe a causal theory of naming, beginning with an initial instance of a baptismal naming of an instance of a natural kind, in this way.) We point to something and say that "dthat" is copper. Then we put things into a set when they are "the same sort of material" as dthat thing. We thus are using the initial hunk of stuff as an exemplar for that set. Some definitions of cardinal numbers work in this way, when we establish

---

23 Fine (2002: 6); (2005: 309).
24 With a property like being 2 or two-ness, the individuals are doubleton sets.

sets based on the equivalence relation of equinumerosity to something taken as the cardinal set, like '{∅}' or '{{∅}, ∅}'.

As a limiting case, propositions can be construed as 0-place relations.[25]

I have given the reduction of monadic to dyadic predicates formally. We might have doubts whether or not this method will work in application; I discuss that issue below.

4. Applications of Predication Theory

In sum, I have proposed a purely formal theory where predication has the basic syntactic structure of a two-place relation. Various configurations of that relation constitute other n-place predicates, both monadic and polyadic.

A formal semantics may be constructed with the same structure. An atomic statement, like 'Rab' is true iff Rab, as Tarski's convention has it. Or, if you like, 'Rab' is true just in case <a, b> is in the extensional R set. Note that 'iff' and 'is in' are also 2-place relations.

But what about applications, interpretations of this predication theory? Is it probable or even meaningful to claim that the world too has truthmakers for these predications consisting in complexes of dyadic relations corresponding to their structure? I end by considering briefly some issues on the ontology.

4.1 The Fit With Frege-Russell Logic

My theory, like the Aristotelian, gives predication a single form. In contrast, Frege's theory does not.

One well known, major difference between Aristotelian logic and modern classical (Frege-Russell) logic lies in their respective treatment of singular and universal predication. Consider statements, one with a singular term as subject term, and the other with a universal term: e.g., 'Socrates is a man' and 'every man is an animal'. In Aristotle's theory, these two have the same logical structure. In Frege-Russell logic they have quite different structures: the first asserts that an individual has the property of being human, as in 'Hs'; the latter asserts that for every individual, if that individual has the property of being human, then it also has the property of being animal, as in '$(x)(Hx \supset Ax)$'. In the latter, the universal proposition makes no assertion about the existence of the subject: as usually construed, '$(x)(Hx \supset Ax)$' is true if no humans exist. But a singular affirmation does require its subject to exist (in 'Socrates is human', 'Hs', 's' must name an element

---

25 Zalta (1983; 2006: 592).

of the domain). So in modern logic 'is' in these two statements functions quite differently and is ambiguous, between 'falling under' and 'being ordered under', as Frege puts it.[26] In contrast, in Aristotle, the predication relation remains the same.[27]

Again, modern logic takes existential statements to have a radically different structure than the predicative ones: 'a human being exists' has only one term, human, existentially quantified: '(∃x)Hx'. It takes 'Socrates exists' to be ill-formed (*'(∃s)s'), or makes up an artificial function, like defining 'Sx' as 'x Socratizes', and then quantifies over it: '(∃x)Sx'.

In contrast, Aristotle shows no indication of breaking up existential and predicative, or singular and universal statements into different logical types. He makes the predication remain always the same, regardless of whether the statement is of *secundum adiacens* ('S is') or of *tertium adiacens* ('S is P'), and whether it be singular or universal. So, unlike modern logic, Aristotle does not recognize explicitly the different logical types. Rather as in the antepredicamental rule, he runs them together [*Cat.* 1b10–5].[28] The modern viewpoint finds proof for the failure of Aristotle's logic in his sanctioning the inference of subalternation from universal to particular affirmative statements, and in his preoccupation with the inference form, 'S is P; therefore S is'.

Yet these disputes do not concern the form of predication that I have advanced: a predication function with two arguments. These different views of predication have that common basis. They disagree in how to symbolize various sentences and in how to classify wffs, with different configurations of that basic structure, into predications of different types.

4.2 The Fit with Natural Language

Linguists claim that predication is an asymmetrical relation between a subject and predicate.[29] My theory of predication does allow for that: there is a relation between the subject and its argument – except that I have two subjects in the basic structure, which may be used to construct such monadic predication. Too, formally, anything may be taken as the subject and anything as the argument: in 'Φx', either 'Φ' or 'x' could be the subject. I even allow, in '(x)(Fx ∨ Ga)" for 'a' to be the predicate and the remainder to be the subject. Such possibilities may be

---

26 Angelelli (1967).
27 Rescher (1978: 58–9).
28 Bäck (2000: Ch. 8).
29 Bowers (1993: 592).

ruled out in an interpretation of the formal structure (what I call a DESCENT) to some natural language, while keeping it as I have it. The standard move in linguistics, of having a natural language as using only fragments of the basic structure, looks similar.

Donna Napoli wants to base predication on semantics and not on syntax: "Predicates are semantical items that need not have any particular syntactic characteristics".[30] But this claim concerns the application of predication theory to natural language. When she says (1989: 4) that "…semantic systems are not isomorphic to syntactic systems", I can agree if she means to claim that the truth-makers for the predications may have quite a different structure.

4.3 The Slingshot Argument

In "True to the Facts", Donald Davidson says that he rejects a traditional correspondence theory of truth where a statement is true if it corresponds to a fact, which fact contains not only the objects that the sentence is about but also what the statement says about those objects.[31] In contrast, Davidson wants to advocate a Tarski-style correspondence theory, where a statement p is true iff p, where 'p' is satisfied by some objects. Thus 'Dolores loves Dagmar' is true given that the ordered pair <Dolores, Dagmar> satisfies the relation 'x loves y'.[32]

I do not object to this approach. That ordered pair is in the set of 2-tuples, the extension of the Love relation. In effect this amounts to A (ii).

The slingshot argument has the form:

R
$\iota x[x = x \& R] = \iota x[x = x]$
$\iota x[x = x] = \iota x[x = x \& S]$
S

where 'R' and 'S' are propositions having the same truth value.

Davidson uses this argument mainly to argue that meaning cannot be "reference", in the Fregean style of *Bedeutung*. As Frege has all true propositions having the same *Bedeutung*, the True, if meaning were "reference", then meaning, then they would all correspond, or "refer", to the same fact. As Gödel remarks, the Slingshot argument assumes Frege's theory and is inconsistent with Russell's, and

---

30 Napoli (1989: 8).
31 Davidson (1969: 749).
32 Davidson (1969: 759).

was already presented by Frege and Church.³³ For 'R' and 'S' would have to be objects in order to be in the identity relation; normally propositions aren't taken thus. Davidson's conclusion would be agreeable to Frege, as for him meaning (*significatio*) would lie with the sense (*Sinn*). Davidson then looks for another way of cashing out meaning, namely in terms of Tarski truth conditions.³⁴

Davidson mixes this point with a general conclusion, that appealing to facts is fuzzy. Just as Quine attacks the notion of synonymy by asking whether 'bachelor' and 'unmarried male' have the same meaning or different meanings, Davidson asks about the individuation of facts: is the cat's being on the mat the same fact or a different fact from the feline's being on the mat, etc. Davidson has a sort of semantic ascent – avoid problems about the world and stay within the way we talk about the world, so as to make the meaning of the sentence, in the context of a particular speech act, is just what the sentence says: the meaning of the cat is on the mat' is that the cat is on the mat etc.³⁵ So: avoid saying that a sentence is true if it corresponds to a fact.

I do not object to that conclusion either, as stated. I do object to inferring from it that a statement can have no truthmaker. What must obtain in the world for Dolores to love Dagmar would constitute its truthmaker. That truthmaker may not resemble the statement much at all: in fact, it would involve two human organisms, each with 10 trillion cells, a nervous system… on a planet such that… at a time when… etc. We do not have a short, nifty way to describe this truth maker with all of its complexities. Still we can at least get at features that it has and ones that it does not have. My writing the previous sentence is not part of the truthmaker for the statement that copper conducts electricity; the "fact" that Cu has 29 protons is though.³⁶

Davidson objects also that dividing up a proposition into subject and predicate gives two entities with no way to hook them up. Hence there is no unity of the proposition.³⁷ Instead Davidson takes the proposition as a whole and uses

---

33 Gödel (1944: 125–53), Davidson (1969: 750; 753 n.6) notes this. Note that to assert 'R = S' is absurd unless you assume, as Frege does, that a proposition names an object.
34 Davidson (1967: 306).
35 Soames (2008).
36 Fine (2014) stresses this point. I am not thinking of a truthmaker as Beebee and Dodd (2005) do when they say when they speak in terms of an entity entailing that a proposition is true. For me a truth maker is not a single entity but a complex, and the relation is not one between propositions and also may be contingent. Perhaps better to speak of grounding, as in Correia and Schneider (2014).
37 Davidson (2006: Ch. 4).

Tarski's method. Here Davidson seems to be making much the same point as Bradley does in his regress argument. But just take predicates as unsaturated to solve this objection.

4.4 The Reduction of Real Properties to the Binary

Above I have offered two ways to reduce properties to dyadic relations. Here I discuss whether such a reduction allows for what we want to say about properties. In short, the second way (B) explains how we learn use terms about properties; the first (A) explains what the sense of a property is.

I have given a reduction of properties via equivalence classes in the manner of Hume and Frege. Here a term is introduced in the context of an equivalence relation, of the form of Hume's Law: if we have a relation '$\Phi(\xi,\zeta)$' that is commutative and associative, then we can write instead of it '$\S\xi, = \S\zeta$'.[38] To use the classic example from the *Grundlagen*: from the relation of equinumerosity, we get: the number of F's ↔ the number of G's iff the F's and the G's are equinumerous.[39] Again, the direction of line L ↔ the direction of line M iff L is parallel to M.[40]

So the monadic property of being a number or of being a direction can be reduced to a 2-place relation. It is introduced contextually in the equivalence relation.

Frege himself says that this type of definition assumes that initially we are to know of the *abstractum*, say, the direction of a line, no other fact but what the relation R asserts, e.g., that it is the same as the direction of another line.[41] The meaning of assertions about the *abstractum* in other respects is being left indeterminate and would require yet further definitions or assumptions. Moreover, we would not be able to say anything about the *abstractum*'s relation to other objects: Frege gives the example of England: the contextual definition leaves it indeterminate whether or not the direction of line a is identical or not to England, as England is not a line at all.

Accordingly Frege disliked and rejected this method of introducing abstract objects because of what has become known as the Caesar Problem. If objects like numbers were introduced in this way, consider the question, 'Is Caesar a number?' That is, is Caesar in one of the equivalence classes of equinumerosity? Caesar is 288 iff Caesar has a one-to-one correspondence to other items in that equivalence

---

38 Frege (1976).
39 Frege (1953: §56); Hume, *Treatise* I.3.1. Cf. Fine (1998: 534). Simons (1998: 487) suggests taking the equivalence '↔' ('iff') as formal equivalence, or maybe something stronger like "synonymy in the object language".
40 Frege (1953: §§64–65).
41 Frege (1953: §65).

class. But Caesar, not being a set, cannot be put into that relation at all. From the way that the abstract object has been introduced via Hume's Law, very few attributes of the objects so defined are known via the equivalence relation. Moreover, just because they cannot be put into that relation at all, objects like Caesar or England have their status as numbers or directions left undetermined.[42]

Thus the Caesar problem is supposed to show that if we use Hume's Law to define and/or introduce abstract objects, we know so little of their attributes that it is silly to introduce them in this way in the first place.[43] Such definitions via equivalence relations look inadequate. (Some neo-Fregeans disagree – at least when we restrict our attention to limited domains, like the domain of numbers.)

In any case, the Caesar problem does not apply to defining predicate functions signifying properties (Frege's "concepts"). Frege himself brings the Caesar problem up for objects. (He takes numbers to be objects, since they serve as the arguments for predicate functions.[44]) But here we are dealing with predicate functions, which signify properties. Given the definition of a property, there is nothing more to know about it itself, except for what is stated in that definition and what follows necessarily from it (what other concepts "are ordered under" [UO] it). To be sure, we can learn about what things ("objects") are in its extension ("fall under [UF] it") and what other properties those things have. But the latter does not concern the property itself.

In traditional terms, the definition of a property yields its intrinsic, essential features; what other features it has extrinsically come about through its being connected to the things having it and their other properties. So the latter are just accidental features of the property.[45]

For instance, consider the property of being a goat. Construct an equivalence relation: (E_1) 'x is the same type as y' or perhaps (E_2) 'x is the same species of animal as y'. To be sure E_1 will give many sets containing some but not all things that we ordinarily call "goats" but mostly not containing only goats. Let Britney and Kevin be goats. They will appear in many sets given by equivalence relations E_1: what we call "goats", "spouses" ('x is the spouse of y'), "American" ('x is the same nationality as y') etc. Yet one such set will be precisely the set of goats. E_1 functions just like the relation of equinumerosity, which generates many sets, each

---

42 Dummett (1981: 402).
43 Resnik (1986: 181) notes "the odd feature of Frege's method" that he does "not assign denotations to abstracts" but "only interprets identities between them".
44 Wright (1983); Heck, Jr. (1993).
45 Haarparanta (1986: 161; 166–167).

of one which is then associated with a distinct cardinal number. Here each one is associated with a distinct property.

Using 'E_2' may be a better, more scientific choice. Once our science progresses, we may get an even better answer, like E_3: 'has the same number of chromosomes as' or 'has the same genetic structure as'. In effect such an equivalence relation expresses the quiddity of a goat: what it is to be a goat. One such set will contain precisely all and only the (normal) goats in the domain. E_3 selects out the right equivalence relations for us.[46]

Again, take copper. 'Copper' is a mass term, and so its instances will be those hunks of material that are pieces of copper. A decent, scientific candidate for an equivalence relation is: 'has atoms having the same number of protons as'. One such set, the 29 proton set, will be precisely the set of pieces of copper. That is, different hunks of matter will end up in different such sets, one of which is the copper set.

What about objects that aren't hunks of material at all, like colors and feelings? These won't satisfy E_2, any such equivalence relation about protons. So they will not end up in any set corresponding to an atomic element. None of them will be copper. Good: so no Caesar problem here.

Take yellowness, Moore's classic example of a simple, irreducible property.[47] 'Yellow' has some ambiguity. We may speak of part of the spectrum of light waves as being yellow. In painting, yellow is a primary color, yet we may speak of certain mixtures of yellow in painting colors as being yellow: yellow ochre, lemon yellow, Indian yellow. In perceiving, we may speak of the *quale* yellow, the yellow of our experience. The latter two senses are often explained in terms of the former: they are given by interaction of certain light waves with the ambient light, reflective surfaces, and the perceptual processes of a (normal, human) observer.[48] Moore focuses on the (experientially) simple property, sc., the *quale* yellow.

The *quale* yellow seems to be a universal: different people can see yellow; I can have a recurring experience of yellow; 'yellow' is a general term in a public, natural

---

46 I do admit that there could be more than one candidate for the equivalence relation defining, say, goat. It is at least logically possible that having the same number of chromosomes (of the right sort) and reproducing its own kind and having macroscopic characteristics have different extensions. This is matter for empirical science: cf. the different ways of defining death, which sometimes gives different results.
47 Campbell (1993: 258).
48 What Moore says (*Principia Ethica* §11) agrees with this. He insists that yellowness is a simple property, yet then admits the experience of yellow may be explained by light waves striking a normal eye etc. Still he says that these are not what we actually perceive, but are what "corresponds in space to what we actually perceive".

language. For 'yellow' taken on its own, without resolution or dissolution, as with 'goat', just use the equivalence relation 'x is the same experience as y': one such set will be the yellow *quale* set. The scientific explanation of the *quale* is that the experience is produced by the causal interaction of a normal human perceiver in normal conditions with light of such a wavelength. So here is a complex of relations, which indeed can serve as truthmaker.

For 'yellow' signifying the property, as a certain predicate of a range of light waves, we can again supply a simple equivalence relation, 'x is the same color as y'. The scientific definition has more interest: x has $Color_n$ iff x has the same wavelength m (could be a range of wavelengths: $c < m < d$) as y. That is, take the latter, which will give equivalence classes. Then name each equivalence class by a name of a $Color_n$, like 'yellow'. So then we associate with each property a 2-place equivalence relation. Then we construct a set of the things in that equivalence relation, and name that set by a name originally associated with a property, like 'yellowness'. Then x is yellow iff x is in that set. Yellowness itself is being in that set; its essence is given by the equivalence relation, more or less satisfactorily, depending on the state of the scientific theory.

For 'yellow' signifying the mental experience, we can start with 'having the same experience as'. As there are many respects in which experiences can be similar, we end up with many equivalence classes. Some of these we might label as color experiences, and of those, as experiences of yellow. To be sure, we don't seem to be able to be too precise here: we may well end up with a very large set of candidate sets for the experience of yellow that we cannot decide between.[49] The particular phenomenon, of the more than usual fuzziness and indecision over which experiences count (especially between people) is more particular to subjective mental experience: lots of indeterminacy here. On account of this, we can see a basis why some people are inclined to reduce mental experiences to their physical bases.

Still, on the less formal, ontological side, one may doubt whether all properties do have such an equivalence relation that is "appropriate" and "satisfactory" – particularly when we have to assume a universal, inflationary domain U. However, given that we admit, *a priori* and/or by necessity, that every property has an extension, which can be characterized as a set, we can stipulate such an equivalence relation. Often we cannot use this relation; the set may not be compressible or compact (in the metalogical sense). (But then the ancients supposed the utility of science or metaphysics to be a defect.)

---

49 This is a general phenomenon common to our knowledge of universals; see my Family Resemblance.

To be sure, (on the epistemological side) clearly we don't always know an adequate, fruitful, and conceptually satisfying equivalence relation for every property. Sometimes we do, as with the example of copper (or one using the conception of wavelengths of light for yellowness). We know this now, and believe that it describes what it is to be a certain color and even what it is to be, or at any rate what it is to produce, the *quale yellow* (when viewed under standard conditions etc.). This one, which has theoretical use and is relatively compressed, took a long time, in the history of science, to get. We should not expect such success for all properties immediately if ever.[50]

Unless we have happened upon a theoretically satisfying and useful equivalence relation, this general reduction of properties to 2-place relations may strike us as formally respectable but pointless in practice. This seems to have been Noonan's intuition, as he insists that the equivalence relation must be is "epistemologically prior" to what is being defined and reduced.[51]

But then we use the second method (B), where a cardinal instance or paradigm is set as one of the *relata* in the equivalence relation. Typically this is how we learn universal terms: the teacher points at an instance, and then we look for other objects similar to it. Eventually, in a natural science like chemistry, a comprehensive theoretical account ends up working better than using exemplars of paradigms – assuming that it is possible. Thus we replace the paradigmatic hunk of copper with an atomic description, and the meter stick at Paris with the amplitude of a wave.[52] Perhaps however this cannot be done for very complex structures, which have no description of a cardinality less than themselves. So Chaitin's Theorem claims.[53] In this case the second method is all we got.

Some, following Peirce, claim that triadic relations, like 'a gave b to c' and structures like Borromean Rings, cannot be reduced to the dyadic.[54] The argument seems to depend on our ability to understand the relations, what they mean. Thus Ketner rejects Quine's method for reducing sequences to dyadic relations, because

---

50 If ever: given inflation (my "exfoliation") or the infinite, or even finitely large (in actual human languages], number of properties or even just the lineation of our scientific research...
51 Noonan (1978: 168–70) says that being the same shape as etc. is "epistemologically prior" to the sortal concepts of direction, shape etc.
52 Cf. Putnam's remarks about water and H_2O in his Twin Earth articles.
53 Chaitin (1980); Casti (1994: 144).
54 Neuman (2014: 35) admits, perhaps inconsistently, that 4-place relations can be reduced to 3-place ones. So too Royce it seems; Russell (1985: 36).

"an act of imagination" is involved.[55] Yet this concerns our ability to understand relations. In any case, it does not rule out relations having the structure that I have advanced. It's just that a relation *in re* may have more content than this formal structure, just as a statement has more content than its predicational form. Perhaps it is better to think of a monadic property as having a relational structure and then emerging from that.[56]

5. Realism

A predicational statement may then be interpreted to have truthmakers in the world. Yet is does not follow that, just because a statement has ultimately a simple, dyadic structure, its truthmaker does too. A statement like 'the banana is yellow (in the sense of seems yellow to me)' may have as its truthmaker a very large complex of factors: photons, surface reflection, ambient light, rod and cone cells, neural networks etc.

Likewise the binary relation of causation, as in 'x is the cause of y', may involve many factors – indeed the whole universe, so some scientists have insisted. Hence they become suspicious of there being any real causal relation. James Clerk Maxwell says:

> It is a metaphysical doctrine that from the same antecedents follow the same consequents. No one can gainsay this. But it is not of much use in a world like this, in which the same antecedents never again concur, and nothing ever happens twice.[57]

Still the theory of predication advanced here does allow for n-place predicates without restrictions, which may have truthmakers themselves constituting a nexus with no upper limit on complexity. It's just that, in empirical fact, as well as in predication theory, such complexity often seems to come from a primary, binary structure, particularly one applied recursively. So I advance a simple theory generating great complexity, and hazard the suggestion that it has *some* isomorphism with the world. We see this pattern in Dawkins' program for the development of the insect wing, in Conway's life game, and above all in the great success of the applications of fractal geometry.

Beyond that, I offer the point that this theory of predication is adequate for expressing all possible statements. As any theory consists in statements, this theory of predication can express all possible statements. After all, all language can be put into binary code. What more can I *say* in its defense?

---

55 Ketner (1995: 270–2).
56 Teller (1992: 142–3).
57 J. C. Maxwell quoted in M. Berry (1978: 111).

References

Asher, N. and Lascarides, A. (2003). *Logics of Conversation*. Cambridge: Cambridge University Press.

Adams, M. McCord. (2000). "Re-reading De grammatico or Anselm's Introduction to Aristotle's Categories", *Documenti e studi sulla tradizione filosofica medievale* 11: 83–112.

Ammonius. (1897). *In Aristotelis De Interpretatione Commentarius*, A. Busse (ed.). *CAG* IV 5. Berlin: F. Meiner.

Angelelli, I. (1967). *Studies on Gottlob Frege and Traditional Philosophy*. Dordrecht, Holland: D. Reidel Publishing Company.

Avicenna (1970). *Al-'Ibāra*. Al-Khudayri, M. (ed.), *Aš-Šhifā*, Part One, Vol. 3. Cairo: M. Dar el-Katib al-'arabi.

Bäck, A. (forthcoming). "Aristotelian Protocol Languages".

Bäck, A. (2000). *Aristotle's Theory of Predication*. New York: E.J. Brill, Leiden-Köln.

Bäck, A. (2015). "Demonstration and Dialectic in Islamic Philosophy". In: L. López-Farjeat and R. Taylor (eds.). *Routledge Companion to Islamic Philosophy*. London & New York: Routledge, 93–104.

Beebee, H. and Dodd, J. (eds.). (2005). *Truthmakers: The Contemporary Debate*. Oxford: Oxford University Press.

Barbour, J. (2000). *The End of Time*. Oxford: Oxford University Press.

Berry, M. (1978). "Regular and Irregular Motion". In: S. Jorna (ed.), *Topics in Nuclear Dynamics*, American Institute of Physics Conference Proceedings 46: 16–120.

Bowers, J. (1993). "The Syntax of Predication". *Linguistic Inquiry* 24: 591–656.

Campbell, J. (1993). "A Simple View of Colour". In: J. Haldane and C. Wright (eds.), *Reality, Representation, and Projection*, Oxford: Oxford University Press, 257–268.

Casti, J. (1994). *Complexification*. New York: Harper Collins.

Chaitin, G. (1980). *Information, Randomness and Incompleteness*, Second Ed. Singapore: World Scientific.

Correia, F. and Schneider, B. (eds.). (2014). *Metaphysical Grounding: Understanding the Structure of Reality*. Cambridge: Cambridge University Press.

Davidson, D. (1967). "Truth and Meaning". *Synthese* 17: 304–323. (Reprinted in Davidson. (2001) *Inquiries into Truth and Interpretation*, 2nd ed. Oxford: Oxford University Press).

Davidson, D. (1969). "True to the Facts". *The Journal of Philosophy* 66 (21): 748–764

Davidson, D. (2006). *Truth and Predication*. Cambridge, MA: Harvard University Press.

Dummett, M. (1981). *The Interpretation of Frege's Philosophy*. Cambridge, MA: Harvard University Press.

Dunlop, D. M. (1958). "Al-Fārābī's Paraphrase of the Categories of Aristotle". *Islamic Quarterly* 4: 168–197.

Dunlop, D.M. (1956). "Al-Fārābī's Introductory Risalah on Logic". *Islamic Quarterly* 3: 224–235.

Fine, K. (1998). "The Limits of Abstraction". In: M. Schirn (ed.), *The Philosophy of Mathematics Today*, Oxford: Oxford University Press, 503–630.

Fine, K. (2002). *The Limits of Abstraction*. Oxford: Oxford University Press.

Fine, K. (2005). "Precis". *Philosophical Studies* 122 (3): 305–15.

Frege, G. (1953). *Grundlagen*. Second Ed. J. L. Austin (ed. & trans.). Oxford: Clarendon Press.

Frege, G. (1976). "Letter to Russell, July 28, 1902". In: *Wissenschaftlicher Briefwechsel*. Hamburg: F Meiner.

Gaskin. R. (2001). "Ockham's Mental Language, Connotation and the Inherence Regress". In: D. Perler (ed.), *Ancient and Medieval Theories of Intentionality*. Leiden: Brill, 227–263.

Gödel, K. (1944). "Russell's Mathematical Logic". In P. A. Schlipp (ed.). *The Philosophy of Bertrand Russell*. Evanston and Chicago: Northwestern University Press, 125–153.

Hale, Bob. (1987). *Abstract Objects*. Oxford: Blackwell.

Heck, Jr., R. (1993). "The Development of Arithmetic in Frege's *Grundgesetze*". *Journal of Symbolic Logic* 58 (2): 579–601.

Haarparanta, L. (1986). "Frege on Existence". In: L. Haaparanta and J. Hintikka (eds.), *Frege Synthesized*, Dordrecht: Reidel, 155–74.

Henry, D. P. (1964). Trans. *The De Grammatico of St. Anselm*. Notre Dame: University of Notre Dame Press.

Ketner, K. (1995). *A Thief of Peirce: The Letters of Kenneth Laine Ketner and Walker Percy*. Jackson, MS: University Press of Mississippi.

Klima, G. (1991). "Latin as a Formal Language: Outlines of a Buridanian Semantics". *Cahiers de l'Institut du Moyen-Âge Grec et Latin, Copenhagen* 61: 78–106.

Lewin, R. (1992). *Complexity*. New York: Macmillan.

Lockwood, M. (2005). *The Labyrinth of Time*. Oxford: Oxford University Press.

Napoli, D.-J. (1989). *Predication Theory: A Case Study for Indexing Theory*. Cambridge: Cambridge University Press.

Noonan, H. (1978). "Count Nouns and Mass Terms". *Analysis* 38 (4): 167–172.

Neuman, Y. (2014). *Introduction to Cultural Computational Psychology*. Cambridge: Cambridge University Press.

Ockham, W. (1970). *Scriptum in Librum Primum Sententiarum Ordinatio*. S. Brown and G. Gál (eds.), *Opera Theologica*. Vol 1. St. Bonaventure, N. Y.: The Franciscan Institute.

Ockham, W. (1974). *Summa Logicae*. P. Boehner, S. Brown and G. Gál (eds.), *Opera Philosophica*. Vol. 1. St. Bonaventure, N.Y.: The Franciscan Institute.

Panaccio, C. (1991). *Les mots, les concepts et les choses*. Montreal-Paris: Bellarmin-Vrin.

Panaccio, C. (2004). *Ockham on Concepts*. Hampshire: Ashgate Publishing.

Quine, W. V. O. (1954) "Reduction to a Dyadic Predicate". *Journal of Symbolic Logic* 19: 180–2.

Quine, W. V. O. (1966). *Methods of Logic*. London: Routledge and Kegan Paul.

Rescher, N. (1978). "The Equivocality of Existence". In: N. Rescher (ed.), *Existence and Ontology*, *American Philosophical Quarterly*, Monograph No. 12: Oxford.

Resnik, M. (1986). "Frege's Proof of Referentiality". In: L. Haaparanta and J. Hintikka (eds.), *Frege Synthesized*, Dordrecht, Reidel, 177–95.

Russell. B. (1985). *Philosophy of Logical Atomism*. La Salle: Open Court.

Simons, P. (1998). "Structure and Abstraction". In: M. Schirn (ed.), *The Philosophy of Mathematics Today*, Oxford: Oxford University Press, 485–501.

Simplicius. (1907). *In Aristotelis Categorias Commentarium*. K. Kalbfleisch (ed.), CAG VIII. Berlin: F. Reimer.

Soames, S. (2008). "Truth and Meaning – In Perspective". *Midwest Studies in Philosophy* 32 (1): 1–19.

Sweeney, E. (2012). *Anselm of Canterbury and the Desire for the Word*. Washington, DC: Catholic University of America Press.

Teller, P. (1992). "A Contemporary Look at Emergence". In: A. Beckermann, H. Flohr and J. Kim (eds.), *Emergence or Reduction?: Essays on the Prospects of Nonreductive Physicalism*, Berlin: De Gruyter, 139–153.

Wright, C. (1983). *Frege's Conception of Numbers as Objects*. Aberdeen: Aberdeen University Press.

Wright, W. (1967). *A Grammar of the Arabic Language Vol. II* (3rd ed.). Cambridge: Cambridge University Press.

Zalta, E. (1983). *Abstract Objects*. Reidel: Dordrecht.

Zalta, E. (2006). "Deriving and Validating Kripkean Claims using the Theory of Abstract Objects". *Nous* 40 (4): 591–622.

Part II.
Philosophy and Beyond

Mieszko Tałasiewicz
University of Warsaw

The Sense of a Predicate[1]

Abstract: The paper examines the notion of the sense of a predicate, as stemming from Frege, in the light of the controversy between Dummett and Geach as to whether the sense of a predicate is a way of getting to the reference of the predicate or rather a function from senses of names to senses of sentences. Several Fregean principles are discussed (Functoriality Principle, Principle of Mediating Sense, Compositionality Principle) as well as the distinction of simple and complex predicates (Dummettian version of this distinction being quite distinct from Fregean one). The conclusion of the paper is that Functoriality Principle can be simultaneously applied with Principle of Mediating Sense, so that the surface form of the Dummett-Geach controversy dissolves. However, Compositionality Principle cannot be held simultaneously with Functoriality Principle, so that there arises a question which principle is worth keeping in the first place. The paper argues that Functoriality Principle should stay, because there is no other sensible way to obtain what it gives us (namely, the unity of the proposition); while there are alternative ways of explaining facts that in Fregean framework were explained through Compositionality Principle (namely, the meanings of complex phrases). Those ways require moving forward from Fregean framework, though, as they essentially involve situation semantics. Un-Fregean as it is, it is a solution.

Keywords: Michael Dummett, Peter Geach, Gottlob Frege, sense, predicate

0. Introduction

The paper is devoted to a critical analysis of the controversy that arose between Michael Dummett and Peter Thomas Geach about the meaning of predicates.[2] Exegetical matters, concerning the actual content of the doctrine of Gottlob Frege, on which both Dummett and Geach commented, will be ignored here.[3] Geach

---

1. This paper elaborates further some ideas originally presented in an article published in Polish as Tałasiewicz (2016).
2. The topic of the present paper largely passes by the topic as Hale (1979). Hale concentrates on the problem of distinguishing subjects from predicates in Strawson, Geach and Dummett, remaining silent on the question of *the sense* of a predicate (as opposed to the *denotation* of a predicate).
3. Ignored will be, too, most of the accusations that Geach and Dummett raised against each other for alleged infidelity towards Frege's thought – except one tasty example: "Dummett's doctrine of simple versus complex predicates […] is a radically un-Fregean

and Dummett worked on Frege because they found in his doctrine elements they themselves regarded as important and valid, and if they felt it justified by the facts of the matter, they both didn't hesitate to correct or improve Frege's solutions. Such is the aim of the present work: to search for the exit rather than contemplate the subtle craftsmanship of the trap.

1. Functoriality *versus* Mediating Sense

According to Fregean Functoriality Principle (FP) a sentence of a certain kind consists always of a complete, saturated name, standing in the subject position and referring to some object, and of an incomplete, unsaturated predicate, **referring to** some concept, *i.e.* to some unsaturated function, an argument of which is the referent of the subject, while the value of which is the referent of the whole sentence (a truth-value).

According to a different Fregean principle – let us dub it the Principle of Mediating Sense (PMS) – expressions do not refer directly, but rather *via* some mediating sense. The sense of a name might be identified as its connotation. The sense of a sentence is a thought expressed in this sentence. But what is the sense of a predicate?

Geach, invoking the idea that Functoriality Principle holds also for the composition of senses[4], insisted that the sense of a predicate – if anything at all can be identified as such a sense[5] – is also incomplete and unsaturated, and takes the form of a function, which to the sense of the subject assigns the sense of the whole sentence, that is the thought expressed in the sentence.[6] Let us add, for the completeness of the account, that the same principle, according to Geach, should be employed at the level of the surface structure of the sentence, too, to the effect

---

    distinction, which cannot be allowed without bringing Frege's fundamental insight into ruinous confusion" (Geach 1975: 150) *versus* "Geach ['s thesis] that we should regard the sense of a predicate as a function mapping senses of names onto thoughts (senses of sentences) […] is inconsistent with Frege's whole conception of sense" (Dummett 1991: 142).

4 "For not all the parts of a thought can be complete; at least one must be 'unsaturated', or predicative; otherwise they would not hold together" (Frege 1892: 54).
5 "I am not concerned, as Dummett is, to defend Frege's doctrine that senses are identifiable objects" (Geach 1975: 150).
6 "What corresponds in the realm of sense to an incomplete expression […] is not an object, a complete sense, but a function with senses as its values and senses as its arguments" (Geach 1975: 150).

that not only senses and denotations of predicates are functions, but the predicates, as expressions, are functions themselves: the so called 'linguistic functions'.[7] Dummett objected. He formulated his objections several times in different contexts using different forms of argument but with the air of rephrasing the same argument. The point of my paper is that all these instances boil down to *two* separate reasons for holding his views against Geach's.[8] In other words, there are two aspects of the problem with (FP) and we need to deal with them separately (even though the protagonists seemed to lump them together).

As to the first reason, appealing to *ratio legis* of (PMS) Dummett claimed that "the sense of the predicate is to be thought of not as being given directly in terms of a mapping from the *senses* of names onto anything (such as thoughts) but, rather, in terms of a mapping of *objects* onto truth-values" (Dummett 1991: 143). "Geach's proposal involves inattention to what we want the notion of sense for" (Dummett 1991: 142), and we want it for proper handling of the *conditions* of truth of a sentence in contrast to the truth itself: knowing the sense of a sentence we know what are its truth-conditions – but not necessarily whether it is actually true or not. And if "to grasp the sense of the predicate is to grasp something that determines its reference, and if it is to lead us to apprehend the condition for a sentence formed by means of that predicate to be true, then it must be given to us as a grasp of the condition that must be satisfied, by any object, for it to be mapped onto the value *true* or for it to be mapped onto the value *false*" (Dummett 1991: 143).

It might be felt, from this reconstruction, that (FP) and (PMS) cannot hold together in the whole range of levels of analysis of compound expressions. In fact it is not a correct conclusion, as both principles can be represented as holding simultaneously at all levels, in the following schema:

---

7   A linguistic function assigns a compound expression, identified as a certain syntactic form, to its argument-expressions. For instance, the functor $\xi * \zeta$ is a two-place function, which for the arguments 2 and 3 yields the value 6; it makes a two-place linguistic function, which for the arguments '2' and '3' yields the value '2*3'. Cf. Anscombe and Geach (1961: 143–144); while the **denotation** of this functor is a two place mathematical function, which for the arguments 2 and 3 yields the value 6.
8   For historical accuracy it might be worth mentioning that Dummett first gave some interpretation of Frege in (1973), to which Geach forcefully (and fiercely) opposed in (1975), formulating his own view, recapitulated above. This started the debate in which Dummett in turn was arguing against Geach.

	subject	predicate	sentence
expression	"The sun"	$f_e: \xi \rightarrow \text{Con}(\xi, \text{"shines"})$	"The sun shines"
sense	connotation of the name "The sun"	$f_s: (\xi, \zeta) \rightarrow$ "ξ shines" is true iff the object satisfying ζ shines $\boxed{\text{PMS}} \quad \boxed{\text{FP}} \rightarrow$	"The sun shines" is true iff the object satisfying the connotation of the name "The sun" shines
reference	The sun	$f_r: \xi \rightarrow 1$ iff ξ shines/ 0 otherwise	1/0

Notes: "Con" stands for concatenation of two expressions. The shape of the predicate reflects its role as a linguistic function. The sense of a the predicate is *de facto* a two-place function, where the first place is kept for the name itself, as an expression, and only the second place is for the actual *sense* of the name.[9]

This schema shows, or so I would argue, that both (FP) and (PMS) can be kept together in the whole range of analysis. We can find such a formula for the sense of a predicate which serves both purposes: assigns thoughts (understood as truth-conditions of sentences) to the connotations of the names and, simultaneously, determines the reference of the predicate, or the function that assigns truth-values to the referents of the names.

That (FP) holds in our schema can be seen directly from the table: the sense of the sentence is the value of the function f_s for the sense of the subject taken as the second argument (the subject itself is to be taken for the first argument). Determination of the reference (PMS) also holds, though perhaps it is not so immediately visible. Let us stop for a moment to show how it goes.

Dummett discusses two interpretations of the thesis that sense determines reference. The weak interpretation, on which "to say that sense determines reference involves only that two expressions with the same sense cannot have different references" (Dummett 1981: 249) is arguably too weak. "On the strong interpretation [...] to grasp the sense is to apprehend the condition for something to be its referent" (Dummett 1981: 249). This interpretation, or so I will argue, is a little bit too strong as a formulation of our common intuitions connected with the notion of determination of reference. Let us consider a simple case of how the connotation of a name determines the name's reference. How, for instance,

---

[9] This complication arises because of the distinction of language and meta-language and has no serious impact on the intuitive grounds of (FP). In certain cases, when the language allows for no strict synonymy between atomic expressions and therefore a one-to-one relation between atomic names and their senses can be established (which means that a name itself might be a value of a function, say function g, taking the sense of this name as its argument), the sense of the predicate still can be a one-place function, such as $f_s: \xi \rightarrow$ ("$g(\xi)$ shines" is true iff the object satisfying ξ shines).

the connotation of the name "The sun" determines its referent: the sun? Well, the connotation of a name is, technically, a set of properties.[10] Is it really sufficient to apprehend this set in order to know what is the condition of being the sun? Of course not. I can perfectly well know this set and have no idea how it constitutes the condition of being the reference of the name "The sun". What I need more is to know some *general* condition of how connotation should be employed to determine reference, namely, apart from knowing some set of properties I must know that *having all properties from this set is necessary and sufficient for being a referent of the name*. Only when we know such a general condition linking the senses of expressions of a certain kind with their references, we can assert that knowing the sense of a particular expression of this kind amounts to apprehending the condition of being the referent of this expression. As for the kind of names, we tend to forget about this general condition as it is plain and obvious. However, such a condition is needed.

For some kinds of expressions other than names this general condition might not be so plain and obvious, after all. Yet it is safe, I presume, to contend that anyway it must be somehow specified, in such a general condition, how the sense of an expression is to be used in specifying the condition of being a referent. In a slightly more depersonalized wording, let us stipulate that the general condition of determination of B by A involves at least that the condition of being B is given in terms available from the formulation of A.

If so, we might depict the determination of the reference of the predicate by its sense in the following way: a referent of a predicate, the sense of which is given as the function $\{f_s: (\xi, \zeta) \rightarrow$ "$\alpha(\xi)$" is true iff α(object satisfying ζ)$\}$, can be characterized as a function assigning the value *true* to all and only such objects ξ, that $\alpha(\xi)$. Thus the sense of a predicate in our formulation determines its reference.

This aspect of Dummett's problem seems to be solved then – in line with the 'functional' intuitions advocated by Geach.

---

10 This example in old Frege's fashion (suggesting a description theory of proper names, which is now rather unpopular) may sound a bit passé, but it still shows the point. First, the subject need not be a proper name; as Geach insists in (1975, 1980) common names are still names (not predicates) and a theory of common names would involve descriptive elements. Second, the example would suffer only in complication, but not in the core point, if amended to the full-blown contextual theory of proper names (provided the theory has not broken up with the idea of a sense of a proper name at all). The sense of a name would be a value of some function from some context (possibly including even the reference of the name). The present argument does not rely on how exactly we construe the sense of a name.

Dummett's insistence that the sense of a predicate cannot be a function is rooted perhaps in a kind of overstatement of Geach's position. "Suppose [as Geach suggests] that the sense of the predicate '*x* stammers' is given to us as a function which carries us from, for example, the sense of the name 'Mrs. Thatcher' to the thought expressed by the sentence 'Mrs. Thatcher stammers'. On Frege's conception of the sense of a sentence, the content of the thought expressed by the sentence depends on the condition for the sentence to be true. What, then, is that condition? What *does* determine the sentence as true or false? On this picture of the sense of the predicate [...] it must be the satisfaction of some condition, given by the predicate, by the *sense* of the name 'Mrs. Thatcher'. The reference of the name will not then come into determination of the truth or falsity of the sentence" (Dummett 1991: 142–143).

That would be implausible, indeed. However – as it is shown above – from the statement that the sense of the name contributes to the truth-condition of the sentence it does not follow that the reference of the name does not contribute to the truth-condition. The sense of the name is taken as the argument of the function identified as the sense of the predicate precisely *qua* the determiner of the reference. In fact, what contributes to the truth-conditions of the sentence is still the reference of the name, as it should be, only that the reference here is identified as something satisfying the sense of the name. In effect we get nothing of the sort that Dummett is afraid of, namely that the truth-conditions of sentences would cease to be linked with factual assignment of truth-values to objects, but rather we get the idea that the truth-condition, taken as the sense of a sentence, or the thought expressed in a sentence, is not detached from the system of the senses of the language, but is expressed with the aid of the meanings available in a given language. In short – when we are talking about truth-conditions, we are still talking about referents of the names, not senses, but about referents *qua* objects satisfying given senses, not *per se*.

Dummett, as it appears, did not discriminate clearly between the reference of the predicate and the sense of the sentence (cf. the table above – the former is in the middle field of the bottom row, the latter is in the right field of the middle row). Both of these can be regarded as determined by the sense of the predicate. Both of these can be called the 'truth-conditions' of the sentence in a certain meaning of the phrase. In both cases it is a different meaning, though, and the cases should not be lumped together. The reference of the predicate is an objective assignment of '0' and '1' to some previously discriminated objects. The thought expressed in a sentence conveys the contribution of the object to truth-conditions *via* the sense of the name. That's why we cannot accept the view that "once the referent of the name has been determined, via its sense, the sense of the name falls away as not further relevant to the determination of the thought as true or false" (Dummett 1981: 252). The predicates "ξ shines" and "ξ emits electromagnetic radiation in visible range" determine the same referent-function (which gives the same assignment of truth-values to objects). Yet the senses of the predicates are different as are the thoughts expressed in the sentences 'The sun shines' and 'The sun emits electromagnetic radiation in visible range'.[11]

2. Functoriality *versus* Compositionality

However, Dummett rejected Geach's views because of some other reason, too. There is another intuitive rule in Frege's stock, namely the Compositionality Principle (CP), according to which the senses of the parts of a compound expression are parts of the sense of this expression. Dummett stresses its great importance: "To say that the sense of the whole is compounded out of the senses of the parts is to say, first, that we understand the complex expression as having the sense it does by understanding its parts and the way they are put together, and, secondly, that we could not grasp that sense without conceiving of it as having just that complexity" (Dummett 1991: 144).

This principle is clearly violated by Geach. It can hardly be denied that "if the sense of '*x* shines' is a function which maps the sense of the phrase 'the sun' onto the thought that the sun shines, then that thought no more contains the sense of 'the sun' as a part than Stockholm contains Sweden as a part" (Dummett 1991: 142).[12]

Now we do have inconsistent principles: we cannot have (CP) and keep (FP) at the level of senses. Which one is more important, or who is right: Dummett or Geach? Well, it depends on what problems are solved by the employment

---

11 Besides, this example – in line with the examples from logic or mathematics – reveals the limitations of the identification of the notion of synonymy with the notion of necessary equivalence. In worlds similar enough to our world the sentences 'The sun shines' and 'The sun emits electromagnetic radiation in visible range' are necessarily equivalent (because shining **is** emitting electromagnetic radiation in visible range) but not synonymous (as 'shining' does not **mean** 'emitting electromagnetic radiation in visible range').

12 Dummett is alluding here to Frege's own example, by which Frege illustrated the thesis that at the level of reference (CP) cannot hold, as the references of the parts of a compound expression aren't parts of the reference of the expression (just as Sweden is not a part of the capital of Sweden), while retaining (CP) at the level of senses.

of respective principles – and whether there are alternative solutions somehow available.

(CP) is a solution to a serious problem of how can we understand atomic sentences. If the senses of predicates were indeed unsaturated functions, as Geach suggests, how could we understand even so easy a sentence as "the sun shines'? Dummett's solution to this problem is based upon a distinction between *simple predicates* and *complex predicates*.[13] Complex predicates are abstracted from sentences: first we must grasp the sense of some atomic sentence, say 'The sun shines', and only then are we in the position to form, through the omission of the occurrence of the name 'the sun', the complex predicate 'ξ shines'. Thus understood, the predicate is needed for a proper account of generality and quantification. It conforms to the Functoriality Principle and instantiates the Fregean notion of unsaturatedness.[14] However, the predicate thus understood cannot in turn be a constituent of the atomic sentence 'The sun shines'. Complex predicates *de facto* aren't expressions at all. They are rather abstract representations of some common features of certain compound expressions.[15]

That is why Dummett goes on to introduce simple predicates, which are identifiable expressions and real constituents of compound sentences, conforming to

---

[13] Dummett (1973: 24–34). Beware of peculiar terminology. Expressions logically simple, according to Dummett, after Frege, are those whose senses are not built-up from some constituent senses. Logically complex expressions are those whose senses, in turn, are built-up in such a way. This has nothing to do with surface structure of the words. Logically simple expressions might be represented by many-word concatenations, while logically complex ones might be reflected in writing only by a single word (accompanied in logical form, but not in surface structure, by certain 'slots'). Dummett himself realized that this terminological convention is a peculiar one, putting on occasion the adjective "complex" between metaphorical quotes, see below.

[14] "Complex predicates form the prototype for Frege's general notion of an 'incomplete' expression. Such expressions are said by him to contain gaps, and, further, to be *unselbständig*: they cannot subsist – they cannot stand up, one might say – on their own" (Dummett 1973: 31).

[15] "Precisely because the 'complex' predicate 'ξ snores' has to be regarded as formed from such a sentence as 'Herbert snores', it cannot itself be one of the ingredients from which 'Herbert snores' was formed, and thus cannot be that whose sense, on Frege's own account, contributes to composing the sense of 'Herbert snores'" (Dummett 1973: 30–31). "The complex predicate 'ξ killed ξ' cannot be regarded as literally *part* of the sentences in which it occurs: it is not a word or string of words, not even a discontinuous string" (Dummett 1973: 31). "The complex predicate is thus not really an expression – a bit of language – in its own right" (Dummett 1973: 31). "A complex predicate is, rather, to be regarded as a *feature* in common to [many] sentences" (Dummett 1973: 31).

Compositionality Principle. In short, apart from a complex predicate 'ξ shines', which is derived from the sentence 'The sun shines', we need also a simple predicate '… shines', which has a self-contained meaning and is a constituent of the sentence.[16]

The idea of distinguishing simple and complex predicates isn't originally Fregean; as Dummett says, Frege 'tacitly assimilated simple predicates to complex ones'.[17] According to Dummett, however, this idea fits the Fregean framework and improves it – or is 'radically un-Fregean', as claimed Geach. Fregean or un-, let us see what this idea gives us, and what it takes from us.

Simple predicates in Dummett's style give us some solution to the problem of graspability of the meaning of atomic sentences – especially if we are ready to accept a popular (but not necessarily best advised) ontological theory according to which properties and relations, on a par with things, are primitive entities (metaphysically or at least epistemologically). On such an assumption we get a semantically natural idea that to specify the truth conditions of an atomic sentence is to specify what property an object should have or in what relations with other objects should it persist. However, if we further assume that this sort of truth conditions is reflected in the syntax of the sentence in such a way that the semantic value of the predicate is such a 'worldly' property or a relation, we are in the risk of effectively turning predicates into a sort of names of properties – which is ruinous for any plausible account of the syntactic role of predicates.[18]

This leads us to the main drawback of Dummett's doctrine of simple predicates. Namely, it takes from us the otherwise available solution to the problem of the unity of proposition, as it is often called. If the sense of a predicate is a function, after all, as Geach persuades, a proposition is a single value of such function – that is why it is unified, being compound. But if the sense of a predicate is just a proper

---

16 "If 'ξ snores' is treated as a complex predicate, on all fours with, say, 'If anyone snores then ξ snores', we do need to recognize the separate existence of the simple predicate '…snores' as well" (Dummett 1973: 30). "Simple predicates are *selbständig* in the way that complex ones are not: they are merely words or strings of words which can quite straightforwardly be written down" (Dummett 1973: 32).

17 Cf. Dummett (1973: 30).

18 It was syntax what mainly motivated Geach's opposition to Dummett, not semantics. As it is well known, Geach was deeply engaged in interpreting ancient and medieval grammars and contributed to the development of modern Categorial Grammar (1970). Frege's doctrine was for him (in a way in which it was not for Dummett) a founding idea of syntax – not only of semantics. And functoriality was the most relevant part of the doctrine.

part of a proposition, on a par with other parts, how can we tell a proposition from a mere list of senses? How can we distinguish the proposition that the cat sits on the mat from the triple <the cat, sitting-on, the mat>?[19] How could we account for the diversity of syntactic and semantic roles of subjects and predicates?[20]

Dummett says very little to answer such questions. About the slots assigned to predicates but not names he says that "in the case of simple predicates, the slots are external to them, while in the case of complex predicates, they are internal" (Dummett 1973: 32). This explains nothing about why these 'external' slots are nevertheless assigned to predicates but not subjects and Geach rightly names this passage as 'remarkably obscure'.[21] Perhaps Dummett thinks that these problems somehow can be solved without the notion of functoriality. He entertains the idea of some restricted notion of incompleteness (as opposed to the general one, involving functoriality), which boils down to the notion of mere logical valency[22] – but this is void unless backed up by some deeper explanation (for it still lefts open the question what makes predicates, but not names, logically 'valent'). Such a backup is available in the literature in the form of Peter F. Strawson's descriptive metaphysics, and indeed a somewhat enigmatic reference to the work of Strawson can be found in Dummett (1973: 32). Strawson soberly acknowledged the charge that Frank Ramsey raised against the notion of such incompleteness that is not based upon functoriality[23] (Strawson 1959: 153), but in spite of that he managed to find such a solution, an un-Fregean one. In search for a proper account of

---

19 Geach explicitly formulated a similar charge in the form of a question of how could we distinguish, upon Dummett's doctrine, the proposition that John hit Mary from the proposition that Mary hit John. Cf. Geach (1976: 444).
20 As to syntactic patterns, what needs explanation is, for instance, that even "to the simple predicates [...] we must assign slots into which singular terms have to be fitted, rather than ascribing to the singular terms slots into which the predicates [...] have to be fitted" (Dummett 1973: 32) The slots assigned to simple predicates are represented in Dummett's notation by three dots, in contrast to the slots contained in complex predicates, represented by Greek letters. On the side of semantics, a good example of phenomenon waiting to be explained is that the negation of a predicate is equivalent to the negation of a respective sentence, while the negation of a subject has no such feature.
21 Geach (1975: 147).
22 "Simple predicates are not literally incomplete", "If incompleteness is ascribed to simple predicates, this can only be as a means of expressing the conception of logical valency" (both citations from Dummett 1973: 63).
23 Roughly: if the predicate is incomplete as a sentence, so is the name. Incompleteness so understood does not differentiate between names and predicates.

syntactic relations he construed a whole metaphysical system in which the role of a name, that is the identification of an object, was depending on acquaintance of the speaker with some empirical fact (reportable by a complete sentence), which was making names something complete, while the role of a predicate, that is the identification of a concept, would not necessitate such liaisons (and therefore could be regarded as incomplete – in this weird but intelligible sense) – cf. (Strawson 1959: 187–188). Strawson in a chain of publications spanning twenty-five years (starting from his celebrated 1950 and culminating in his 1974, of which he himself said that it was his most ambitious book, but the one that had received the least attention) developed a very sophisticated syntactic theory. But the theory collapsed, eventually, as a correct account of syntax, or so I have argued elsewhere.[24] And even if it didn't, Strawson's solution is hardly a widely accepted one and it is doubtful whether Dummett would be ready to accept the price for it by accepting Strawson's peculiar metaphysical system.

3. The Way Out: Situation Semantics

Thus we are near the point of departure of section two. Instead of asking which principle is more important we might ask which problem is in a greater need of being solved. Or – which can be solved in some other way.

An attempt to harmonize Geach's and Dummett's stance is made in (Simons 1981). I do not think, however, that it is successful. Simons tries to divide the merits, saying that "on grounds of mere simplicity, Geach wins, since Dummett has two senses [simple and complex], one of which is also Geach's [complex]. But Dummett is also right that, in any language richer than the austere Tractarian one, atomic sencences will contain expression other than names, which, when put together with names according to syntactic rule, yield sentences" (Dummett 1973: 94). Effectively, Simons takes Dummett's side, as he shares Dummett's view that "the attempt to list predicates without bringing in rules for forming atomic sentences, out of verbs and names, necessarily fails" (Dummett 1973: 95). This is precisely what Geach is opposed to.

The solution I am going to offer in the present paper is intended to keep to the basic idea that atomic sentences are prior to formation of predicates, but reject the notion of forming atomic sentences out of verbs and names. We can have full Functoriality Principle at all levels, involving the unsaturatedness of predicates, accommodate Dummett's observation that unsaturated predicates can be

---

24 Tałasiewicz, forthcoming

obtained only as abstractions from prior sentences and account for graspability of the sense of atomic sentences. All that we need is to take (some) atomic sentences as logically simple expressions – such that their sense is not calculated by any composition, but is given directly.

This solution is by no means cheap. It requires quite substantial underlying assumptions. In the first place, the assumption that some sentences can be defined ostensively. That in turn requires some situation semantics, and not just any sort, but such that at least some of the situations are visible particulars, ready to be pointed at without any need of comment or abstraction. Otherwise this solution would suffer from the charges raised by Dummett, who has considered a similar account but unsupported by situation semantics (Dummett 1973: 293–294).[25]

According to the first of these charges, the truth-condition of a sentence must always be construed as some condition that should be satisfied by the referent of the subject of the sentence. If so – and Geach is inclined to give his consent here[26] – Dummett is ready to show that we need to have some sense specifying this condition, other than the sense of the subject, before we understand the sentence. And that is precisely the sense of a simple predicate. Only situation semantics, which gives us the truth-condition of a sentence in terms of some situation's being a fact (which can be realized without having individuated the referents of the names within this situation), can absolve us from this charge.

The second charge, if sound, would hit even the situation semantics. According to that charge even if we allowed demonstratively introduced sentences, they would be hidden incomplete expressions, namely ones which contain a significant present tense, and therefore could not serve as basis for further abstraction of complex predicates. Fortunately, this charge isn't sound. First, it is not at all clear why all demonstratively introduced sentences must contain a significant present tense. The act of demonstration of course has a definite temporal index, but it does not follow that it must be inevitably incorporated into the sentence. Dummett thinks that "it is impossible to conceive of any method of grasping [demonstratively] the thought expressed by a sentence like 'The Earth is round'" (Dummett 1973: 294). Indeed, the meaning of this sentence is hardly demonstrable, but because of the size of the Earth rather than for its alleged status of an "eternal truth" or the lack of significant tense. 'This orange

---

25 Similar charges are rephrased in a lengthy form, entangled with exegetical matters, in (Dummett 1981, Chapter 16, particularly p. 294).
26 "It is clear that [...] to understand a sentence we must understand the names occurring in it" (Geach 1976: 442).

is round' is quite conceivable, isn't it? Furthermore, suppose that a sentence has a significant present tense, after all. Why would it make this sentence an incomplete expression? It makes it a context-dependent expression, true, but in the context of an utterance such a sentence is a complete expression, which cannot be said of any (complex) predicate. Present tense in an utterance is a *name* of a moment of time, not a *slot* for inserting one.

Situation semantics is a serious commitment, leading us beyond the legacy of Frege, but – unlike the even more committing ontology devised by Strawson – it has a rich tradition behind it and manifold independent applications. The tradition reflects itself in the work of many prominent philosophers since Wittgenstein, but a milestone of it, and a turning point from which it became a really powerful tool in analyzing natural language, was of course *Situations and Attitudes* by Barwise and Perry: "We are always in situations; *we see them*, cause them to come about, and have attitudes toward them" (Barwise, Perry 1983: 7; italics mine).

The details of the proposed solution, revealing the exact syntactic mechanisms, responsible for the construction of propositions – and in general the reconstruction of grammar from some foundational philosophical assumptions – are presented in my *Philosophy of Syntax* (Tałasiewicz 2010: 88–115). It might be worthwhile, though, to highlight several points particularly relevant to the problem of the sense of a predicate. The first important point is that in the ontology of situations – in contrast to Strawson's ontology of objects and concepts, for instance – properties and relations are not the ultimate "bricks" the world is built of; they are rather 'uniformities across situations' (as Barwise and Perry would have it) or similarity classes of situations. There is no need to postulate a special class of expressions referring to them and to identify this class with the class of predicables.[27] Self-contained senses can be assigned only to names – and whole sentences. It reflects some old intuition, expressed by Husserl, that the two basic semantic categories of names and sentences correspond to the two basic forms of

---

27 I owe this excellent terminological invention to Geach. In his (1980) he noticed that while traditional terminology allows for distinguishing the syntactic position of the subject and the semantic category of the name that is *capable of* standing in this position, it fails to provide analogous resources to distinguish syntactic position of the predicate from respective semantic category of expressions *capable of* standing in this position. To this end he introduced the category of *predicables*. Thanks to this, we now have in our possession dual terms at both levels: subject-predicate and name-predicable.

intentional acts: nominal and propositional.[28, 29] Properties and relations, taken as abstractions from situations, can be of course referred to – by naming them and forming judgments about them. There is no particular link, however, between these kinds of abstract objects on one side and the class of predicables on the other. An expression that refers to a property or a relation is a name of this property or relation, not a predicable. Predicables standing in predicate positions are structural, syncategorematic signs that serve the purpose of building new sentences (the senses of which were not demonstrated) or representing *a posteriori* the structure of sentences defined demonstratively, as it is reflected upon in subsequent syntactical analysis. Their senses are by no means graspable on their own, and might well be given, conventionally, as functions – such as depicted in our table.

References

Ajdukiewicz, K. (1967). "Syntactic Connexion". In: S. McCall (ed.), *Polish Logic 1920–1939*, Oxford: Oxford University Press, 207–231.

Anscombe, G. E. M. and Geach, P. T. (1961), *Three Philosophers*. Ithaca, NY: Cornell University Press.

Barwise, J. and Perry, J. (1983). *Situations and Attitudes*. Cambridge, MA: MIT Press.

Black, M. *and* Geach, P. T. *(eds.) (1960). Translations from Philosophical Writings of Gottlob Frege*. Oxford: Basil Blackwell.

---

28 "Nominal acts and complete judgments never can have the same intentional essence, and [...] every *switch* from one function to the other, though preserving communities, necessarily works changes in this essence" (Husserl 1970: 152). "Naming and asserting do not merely differ grammatically, but 'in essence', which means that the acts which confer or fulfil meaning for each, differ in *intentional essence*" (Husserl 1970: 158). "Among nominal acts we distinguish positing from non-positing acts. The former were after a fashion existence-meanings [...] refer to [an object] as *existent*. The other acts leave the existence of their object unsettled: the object may, objectively considered, exist, but it is not referred to as existent in them, it does not *count* as actual, but rather 'merely presented'. [...] We find exactly the same modification in the case of judgements. Each judgement has its modified form, an act which merely presents what the judgement takes to be true [...] without a decision as to truth and falsity [...]. Judgements as *positing propositional acts* have therefore their merely presentative correlates in *non-positing propositional acts*" (Husserl 1970: 159–160).

29 Husserlian intuitions were shared and formalized into the form of Categorial Grammar by Stanisław Leśniewski and Kazimierz Ajdukiewicz (cf. Ajdukiewicz 1967).

Dummett, M. (1973). *Frege. Philosophy of Language*. Cambridge MA: Harvard University Press.

Dummett, M. (1981). *The Interpretation of Frege's Philosophy*. Cambridge, MA: Harvard University Press.

Dummett, M. (1991), *The Logical Basis of Metaphysics*. Cambridge, MA: Harvard University Press.

Frege, G. (1891). "Function and Concept". In: M. Black and P. T. Geach (eds.), *(1960), 21–41*.

Frege, G. (1892). "On Concept and Object". In: M. Black and P. T. Geach (eds.), *(1960), 42–55*.

Geach, P. T. (1970). "A Program for Syntax". *Synthese* 22: 3–17.

Geach, P. T. (1975). "Names and Identity". In: S. Guttenplan (ed.), *Mind and Language*, Oxford: Clarendon Press, 139–158.

Geach, P. T. (1976). "Critical Notice. Frege: Philosophy of Language. By Michael Dummett etc". *Mind* 85: 436–449.

Geach, P. T. (1980). *Reference and Generality*. Ithaca, NY: Cornell University Press.

Hale, B. (1979). "Strawson, Geach and Dummett on Singular Terms and Predicates". *Synthese* 42: 275–295.

Husserl, E. (1970). *Logical Investigations*. Vol. 2. London: Routlege.

Simons, P. (1981). "Unsaturatedness". *Grazer Philosophische Studien* 14: 73–95.

Strawson, P. F. (1950). "On Referring". *Mind* 59: 320–344.

Strawson, P. F. (1959). *Individuals*. London: Routlege.

Strawson, P. F. (1974). *Subject and Predicate in Logic and Grammar*. London: Ashgate.

Tałasiewicz, M. (2010). *Philosophy of Syntax. Foundational Topics*. Dordrecht: Springer.

Tałasiewicz, M. (2016). "Spór o sens orzeczenia". In: A. Brożek et al. (ed.), *Myśli o języku, nauce i wartościach. Seria druga. Prof. Jackowi Juliuszowi Jadackiemu w 70. rocznicę urodzin*, Wydawnictwo Naukowe Semper, Warszawa 2016, pp. 128–137.

Tałasiewicz, M. (forthcoming). Subject and Predicate. A Critical Note on Peter F. Strawson's Philosophy of Grammar.

Danny Frederick
Boston, United Kingdom

The Unsatisfactoriness of Unsaturatedness

Abstract: Frege proposed his doctrine of unsaturatedness as a solution to the problems of the unity of the proposition and the unity of the sentence. I show that Frege's theory is mystical, ad hoc, ineffective, paradoxical and entails that singular terms cannot be predicates. I explain the traditional solution to the problem of the unity of the sentence, as expounded by Mill, which invokes a syncategorematic sign of predication and the connotation and denotation of terms. I streamline this solution, bring it up to date and contrast the resulting conventionalist account with Frege's unsaturatedness account. I argue that the conventionalist account provides a clear and intelligible solution to the problem of the unity of the sentence which is free of the defects of Frege's account. I suggest that the problem of the unity of the proposition is spurious. I recommend that the notion of unsaturatedness be extruded from serious debate.

Keywords: conventionalist, Gottlob Frege, John Stuart Mill, unity of the proposition, unity of the sentence, unsaturatedness

1. Introduction

A proposition is the content of a sentence that may be true or false. It may be expressed by different sentences in the same or different languages. A simple form of proposition is the atomic proposition, in which, to put it loosely, an individual is characterised in some way, or in which one or more individuals are specified to be in some relation to each other.[1] Atomic propositions are expressed by atomic sentences, such as the following:

(1) Socrates is wise
(2) John dislikes Peter

In (1) we have one singular term, 'Socrates', and a one-place predicate, 'wise'. In (2) we have two singular terms, 'John' and 'Peter', and a two-place predicate, 'dislikes'. Sometimes a singular term occurs more than once in the same atomic sentence, as in 'John dislikes John' (where both occurrences of 'John' denote the same man). We can distinguish between predicates according to the number of occurrences of singular terms they require to form an atomic sentence. We thus

---

1 A precise characterisation is given in section 3.

distinguish between monadic predicates (like 'wise') and relational predicates; and among relational predicates we can distinguish those which are dyadic (like 'dislikes'), those which are triadic (like 'in between'), and so on.

The problem of the unity of the proposition is to explain how the elements that make up a proposition can be combined in such a way as to produce something that may be true or false. The elements of the proposition expressed by (1) may appear to be Socrates and wisdom, while the elements of the proposition expressed by (2) may appear to be John, dislike and Peter. But a collection of elements does not amount to a proposition that may be true or false. Something has been left out. We might try invoking a relation such as exemplification or instantiation. For example, we might analyse the proposition expressed by (1) into Socrates, wisdom and instantiation. But this just gives us three separate elements instead of two: the unified proposition still eludes us. An attempt to invoke some further relation, for example, to relate Socrates, wisdom and instantiation, will similarly add another propositional element without producing the unified proposition that may be true or false; and so on ad infinitum. It seems that any attempt to analyse a proposition only identifies its elements and loses the bond that makes them into a proposition.

The problem of the unity of the sentence is to explain how expressions that stand for things can be combined in such a way that they make a sentence that says something that may be true or false (that expresses a proposition). If we analyse the sentence, (1), into expressions that stand for Socrates and wisdom, we convert the sentence into a two-term list. If, to try to circumvent this, we analyse (1) into expressions that stand for Socrates, wisdom and instantiation, we convert the sentence into a three-term list; and so on. The problem seems to be parallel to the problem of the unity of the proposition.

In section 2, I expound Frege's 'unsaturatedness' solution to both problems. In section 3, I recount the solution to the problem of the unity of the sentence that was offered by traditional logicians and I develop it into what I call the 'conventionalist' account. In section 4, I show that, while the conventionalist approach gives a satisfactory solution to the problem of the unity of the sentence, Frege's unsaturatedness account gives a satisfactory solution to neither of the problems of unity because it is mystical, paradoxical, ad hoc, devoid of explanatory value and has untoward consequences. In section 5, I comment briefly on some contemporary philosophers who have taken faltering steps toward the conventionalist account. In section 6, I suggest that the problem of the unity of the proposition is a pseudo-problem. In section 7, I sum up and I recommend that Frege's notion of unsaturatedness should be jettisoned.

2. Unsaturatedness

Frege's proposed solution to the problems of propositional and sentential unity invokes the notion of unsaturatedness or incompleteness. On this view, propositions contain parts that are unsaturated or essentially incomplete which, when conjoined with appropriate other parts, form a propositional unity; and different types of linguistic expression correspond to different parts of a proposition; so when appropriate types of linguistic expression are combined they form a sentence.

> For not all the parts of a thought [that is, a proposition] can be complete; at least one must be 'unsaturated,' or predicative; otherwise they would not hold together. For example, the sense of the phrase 'the number 2' does not hold together with that of the expression 'the concept *prime number*' without a link. We apply such a link in the sentence 'the number 2 falls under the concept *prime number*'; it is contained in the words 'falls under,' which need to be completed in two ways – by a subject and an accusative; and only because their sense is thus 'unsaturated' are they capable of serving as a link. Only when they have been supplemented in this twofold respect do we get a complete sense, a thought (Frege 1892a: 54).

Thus, in an atomic sentence, both singular term and predicate have a sense that is incomplete in that it does not expresses a proposition; but the sense of the predicate is in a *further* way incomplete, or unsaturated, in contrast to the complete or saturated sense of a singular term. The unsaturatedness of the sense of the predicate is such that, when the predicate is conjoined with the appropriate number of singular terms, all the senses interlock to form a proposition which is expressed by the resulting sentence. In line with this theory, and in contrast to traditional logicians, Frege sometimes counts the copula as being part of the predicate or 'concept-word' (1892a: 46, 48; 1903: 569; 1906a: 177–178, 180; 1914: 237, 240). At other times he conforms to traditional usage (1892a: 43–44; 1897: 126–38; 1914: 233; 1915: 251–252). So, the reason that the analysis of the sentence (1) into terms standing for Socrates and for wisdom ceases to express a proposition is that we have replaced the predicate, 'wise' or 'is wise', which has an unsaturated sense, with the singular term 'wisdom', which has a saturated sense and so does not hold together with the sense of 'Socrates' to form a proposition.

Frege distinguished simple predicates, like 'wise' or 'is wise', from complex predicates that are formed *from* sentences by the removal of one or more occurrences of a singular term. The sentence from which the complex predicate is formed may be atomic, compound or quantified, as in the following examples of complex predicates:

(i) …is taller than John;
(ii) If Peter is taller than…, then…dislikes Peter;
(iii) Everyone, y, is such that, if y is taller than…, then…dislikes y.

Such *complex* predicates are literally incomplete, *syntactically*, being abstractions from sentences. They can be viewed as sentences with holes punched in them. Further, it seems plausible that the only way we could come to understand the meaning of such a complex predicate is by first understanding a sentence in which it can be discerned and then abstracting away the relevant occurrences of singular terms. In contrast, *simple* predicates, like 'taller than', or 'is taller than', and 'dislikes', are among the elementary expressions from which atomic sentences are formed: they are not syntactically incomplete; and our understanding of the meaning of atomic sentences, like (1) and (2), depends upon our understanding of the meaning of their constituent simple predicates, such as 'wise', or 'is wise', and 'dislikes'. However, on Frege's theory, both simple and complex predicates have an unsaturated sense because that is required to solve the problems of the unity of the sentence and the unity of the proposition (Dummett 1981a: 27–33; 1981b: 261–322).

Since predication is not the only kind of propositional combination, it is not only predicates that are unsaturated. The logical connectives and quantifier phrases have unsaturated senses and stand for incomplete entities, namely, functions of different kinds. For instance, 'if' on Frege's view, is an unsaturated expression joining two sentences into a compound sentence in which one sentence is antecedent and the other is consequent.

In contemporary logical theory it has become standard, following Frege's intermittent practice (inspired by his unsaturatedness doctrine), to regard the copula as part of the predicate, and even to assimilate simple predicates to complex predicates, as Frege himself had done (1891: 31), for example, by writing '…is wise' in place of 'is wise'. This makes sense if one accepts Frege's theory of unsaturatedness. However, it would be question-begging to argue against alternatives to the unsaturatedness theory by insisting that the copula, and even an ellipsis, must form part of the predicate.

3. Convention

Traditional logicians solved the problem of the unity of the sentence by distinguishing between *terms* (categoremata), which have meaning by standing for things, and *syncategoremata*, which have meaning without standing for things (MacFarlane 2009, section 1). John Stuart Mill, expounding a version of the traditional view, says:

> A predicate and a subject are all that is necessarily required to make up a proposition: but as we cannot conclude from merely seeing two names put together, that they are a

predicate and a subject, that is, that one of them is intended to be affirmed or denied of the other, it is necessary that there should be some mode or form of indicating that such is the intention; some sign to distinguish a predication from any other kind of discourse (Mill 1843, book 1, chapter 4, section 1, 125).

He goes on immediately to say that the sign of predication is the copula, or the verbal form that is given to the predicate in the absence of the copula (see also Arnauld and Nicole 1662, part 2, chapter 2, 104, 106–107).

Mill says that a general term *connotes* a property and *denotes* the things (if any) that exemplify that property; while a proper name *denotes* the thing it names and *connotes* nothing (1843, book 1, chapter 2, section 5, 94–96). Thus, in (1), the general term 'wise' connotes the property of wisdom and denotes each wise thing (each thing that exemplifies the property), while the singular term 'Socrates' denotes Socrates, who is one of the things exemplifying wisdom. The meaning of 'is' in (1) can therefore be explained by means of the following convention, which assigns the truth-condition of the sentence:

> the concatenation of the singular term 'Socrates', 'is' and the term 'wise', in that order, is true if and only if what 'Socrates' denotes exemplifies what 'wise' connotes (compare Mill 1843, book 1, chapter 5, section 4, 138–39).

I propose four modifications of this traditional account in order to streamline it and bring it up to date. The modified account will be labelled 'conventionalist'.

First, Mill's formulation mistakenly identifies predication with affirmation or denial; but, as Frege (1906b: 185–186) explained, one may propound a proposition by means of an indicative sentence without asserting or affirming it. For instance, one may employ the sentence 'Socrates is wise' in merely supposing that Socrates is wise. Or one may assert the conditional, 'If Socrates is wise, Plato is wise', without thereby asserting or affirming its antecedent. It is easy, though, to amend the account to accommodate this point. One maintains that (in English and many other languages) the form of the verb is the sign, not of assertion, but of predicative combination: it signifies the combination of terms into the expression for a proposition.

Second, the account needs to accommodate Frege's insight that atomic sentences may be relational. Third, Mill's denial of connotation to proper names introduces an unnecessary complication into his account. We take cognisance of these two points by explaining terms in the following way. Every term connotes a property. A non-relational term denotes each thing that exemplifies the property it connotes; a dyadic relational term denotes each ordered pair of things that exemplify the relational property it connotes; a triadic relational term denotes each ordered triple of things that exemplify the relational property it connotes;

and so on. A *singular term* is a non-relational term that connotes a property that can be exemplified by at most one thing and at least one thing. Thus, it has a unique denotation if it has any denotation; but, unlike those general terms such as 'human and not human,' which denote at most one thing because they cannot denote anything, a singular term is such that it is possible that it denotes. Therefore, in (1), 'Socrates' *connotes* the property, *being Socrates*, or *Socrateity*, and it thereby *denotes* Socrates, and 'wise' *connotes* the property, *wisdom*, and it thereby *denotes* each wise thing. We therefore have two senses in which a term stands for an entity. Every term stands for a property, in the sense of connoting it. Every singular term which denotes a thing stands for that thing in the sense of denoting it. Since it is awkward to say that a general term stands for each of the many things it denotes, we say that only singular terms stand for things in the sense of denoting them.

Fourth, we must take account of Frege's distinction between simple and complex predicates. Thus, we count each complex predicate as a complex term, our understanding of which is dependent upon our understanding of a sentence in which it may be discerned. Although such terms are syntactically incomplete, we can reformulate them as syntactically complete expressions, for example, by prefixing them with 'a thing such that' and then filling all the gaps with an appropriate pronoun (or, in the case of a dyadic complex predicate, 'a pair of things such that,' and so on), or by filling the gaps with free variables to produce what, in contemporary logic, is called an 'open sentence.' Recognising the existence of complex terms does not involve a commitment to the existence of unsaturated senses. Even a syntactically incomplete complex term connotes a property, just as a simple term does. For example, the three syntactically incomplete complex general terms, (i)–(iii), listed in section 2, *connote*, respectively, the three properties:

- being taller than John;
- disliking Peter if Peter is taller than he/she is;
- disliking everyone who is taller than him/her.

It should be noticed that we just *denoted* the properties by means of noun-phrases; we could have used definite descriptions ('the property of being taller than John,' and so on); and we could even give the properties proper names (the significance of this point will become clear in sub-section 4.3, below).

The sign of predicative combination can be seen in any atomic sentence, so we offer the following convention as a general explanation of its meaning:

(C1) An ordered set of an *n*-adic term, *n* occurrences of *m* singular terms (*n* ≥ *m*), and the sign of predicative combination is a sentence which is true if and only if the things denoted by the *n* occurrences of the singular terms, taken in order, exemplify the property connoted by the *n*-adic term.

The *m* singular terms, in each of the *n* occurrences, are the *subjects*; the *n*-adic term is the *predicate*. The property connoted by the predicate is *predicated of* the things denoted by the subjects. The sign of predication is not a term: it does not connote a property and it does not denote (it is syncategorematic). Its function is to indicate that the occurrence of the *n*-place term and the *n* occurrences of the singular terms (in order) are combined into an expression which is true *if and only if* the things denoted by the *n* occurrences of the singular terms, taken in order, exemplify the property connoted by the *n*-place term.

On this view, the problem of the unity of the sentence arises if we mistakenly assume that an atomic sentence can be analysed without remainder into a set of terms, that is, into a set of expressions that stand for entities, either by connoting or by denoting them. The mistake is corrected when we notice that an indispensable element in any sentence is a sign that is not a term, that is, a sign the conventional function of which is not to stand for something but is rather to symbolise the combination of terms into a sentence. This sign, in an atomic sentence, is the sign of predication. It does not connote the property of predication or any other property, and it does not denote anything. Its function is to *symbolise* the predicative combination of terms; but it is not itself a term.

Predication is not the only kind of propositional combination. The logical connectives (conjunction, disjunction, and so on) and quantifier phrases (perhaps 'Everything, *x*, is such that' and 'Something, *y*, is such that') symbolise other forms of propositional combination, without standing for them; and they are also explained in terms of the truth-conditions of the resulting sentences. For example, the meaning conventionally assigned to 'and' is such that, when we place a sentence on either side of it, what results is not a list of two sentences and an expression connoting or denoting conjunction but, rather, a sentence which is true if and only if each of the two conjoined sentences is true.

The account given is purely 'logical' or abstract: different languages will realise the semantical features in different ways. For example, the general explanation of predication refers to an ordered set of expressions; but languages will differ as to which orders are grammatical. Indeed, as we noted, verbs may combine the predicative sign with a term (as 'lives' is equivalent to 'is living'). Further, in some languages a single expression may convey a whole atomic sentence, thereby combining in itself two terms and the sign of predicative combination, as 'vivo' in Latin means 'I live', or as a single word may be used as code for a sentence. We can also imagine languages without verbs, in which the predicative sign takes another form (concatenation, for example, as happens in Russian and in Arabic, in some contexts). Variation in grammars does not detract from the underlying unity of logical function.

4. Comparative Evaluation

In the following three sub-sections I show that the unsaturatedness solution to the problems of the unity of the sentence and of the proposition has some serious defects from which the conventionalist solution to the problem of the unity of the sentence is free.

4.1 Explanation

The Fregean solution to the unity problems is mystical and ad hoc. It is mystical because the property of unsaturatedness is mysterious. Frege himself admits that his talk of completeness and unsaturatedness is only figurative, and that he is merely giving hints (1892a: 55). The solution is ad hoc because the property of unsaturatedness is factitious. The different elements of the proposition combine to form a unity because one of the elements is incomplete; but to say that one of the elements is incomplete is to say no more than that it is so made as to combine with the other elements in such a way as to form a unity that is a proposition. Different terms combine to form a sentence that expresses a proposition because the sense of one of the terms is unsaturated; but to say that the sense of one of the terms is unsaturated is to say no more than that it is so made as to combine with the senses of the other terms in such a way as to form a sentence that expresses a proposition. That is vacuous. It is on a par with Molière's explanation that opium causes sleep because it has a 'dormitive virtue'. The supposed explanation merely replaces one mystery with another.[2]

Indeed, it is worse than that, because it is also ineffective. For, we can take expressions which supposedly have saturated and unsaturated senses and make them into a list: the postulation of unsaturatedness does not explain the difference between such a list and a sentence. Similarly, it is not clear what distinguishes a proposition from a collection of saturated and unsaturated senses (MacBride

---

2   Molière was ridiculing the late-mediaeval scholastic philosophers whose explanations were regarded as vacuous and 'occult' by the mechanical philosophers who heralded the rise of modern science (see, for example, Roux 2013: 71–80).

2005: 605). In consequence, the Fregean solution to the problems of the unity of the proposition and of the sentence is utterly devoid of explanatory value.

It might seem that the conventionalist solution to the problem of the unity of the sentence lacks explanatory value in a similar way. It says that different expressions combine to form a sentence that expresses a proposition because one of the expressions has the function of symbolising predication without standing for anything; but to say that one of the expressions has the function of symbolising predication without standing for anything is to say no more than that this expression is so made as to combine with the other expressions in such a way as to form a sentence that expresses a proposition. However, there is nothing *mysterious* about an expression being so made as to have that function. It is a commonplace that we can, and do, give expressions the functions they have by means of conventions. For example, 'Socrates' has the function of standing for Socrates because we conventionally allocate that sign to that purpose. Further, the explanation of the unity of the sentence in terms of linguistic conventions is not ad hoc, because we appeal to linguistic conventions generally to explain how expressions have the meanings they have. Even Frege must appeal to linguistic conventions to explain how particular expressions have the meanings they have. The conventionalist theory employs the same resources to explain sentential unity, whereas the Fregean theory invokes something extra (and mysterious) purely for that purpose. So, even if Frege's theory could explain sentential unity (which, as we have seen, it cannot), the conventionalist theory would still be preferable on grounds of conceptual simplicity or economy.

In addition, the conventionalist account is superior to Frege's because it can explain how we can make a list by using expressions which, on the Fregean account, are supposed to weld into a sentence if juxtaposed. For, while prevailing linguistic conventions form a background for our uses of language, we can utilise an existing linguistic convention in an unusual way, or set it to one side, or introduce a new convention to override or supplement it. John may refer to a stupid or unreflective person as 'Socrates', and Joy may understand John's ironic remark if she realises that 'Socrates' is conventionally used to denote the wise philosopher. The singular term 'Rhodesia' was conventionally assigned the function of denoting a particular country; but that did not prevent people from renaming that country 'Zimbabwe'. The general term 'mouse' is conventionally assigned the function of denoting small rodents of a particular kind, but it has also been given the function, in some contexts, of denoting pointing devices used with computers. The predicative sign has been conventionally introduced to effect the predicative combination of other expressions, the relevant convention being specified by (C1), and we normally avail ourselves of that convention to utter sentences; but we may sometimes bypass or override it. Sheila might utter the words 'Socrates', 'is' and 'wise', in that order, intending to utter a list of words rather than a sentence; and, given sufficient contextual clues, her audience may understand that she is making a list rather than uttering a sentence. Bill could introduce a new convention that, in specific contexts, an expression which normally is an atomic sentence, is to be understood as a list of expressions consisting of a singular term and an expression combining the copula with a term. Conventions are malleable in ways that unsaturatedness is not.

4.2 Singular Terms

A direct consequence of Frege's doctrine of unsaturatedness is that singular terms cannot be predicates. If it is the unsaturatedness of the sense of the predicate that is responsible for propositional unity, then replacing the predicate with a singular term, which has a saturated sense, will transform a sentence into a list. The problem with this consequence is that it seems to be false. For example, in (1), if the predicate, 'wise', is replaced by the singular term, 'Socrates', what is obtained is not a list but a true sentence, namely:

(3) Socrates is Socrates.

Frege notices this problem for his account and he responds by claiming that 'is' is ambiguous as between predication and identity (1892a: 43–44):

> Surely one can just as well assert of a thing that it is Alexander the Great, or is the number four, or is the planet Venus, as that it is green or is a mammal? If anybody thinks this, he is not distinguishing the uses of the word 'is'. In the last two examples it serves as a copula, as a mere verbal sign of predication…We are here saying that something falls under a concept, and the grammatical predicate means this concept. In the first three examples, on the other hand, 'is' is used like the equals sign in arithmetic, to express an equation.

So, when we substitute 'Socrates' for 'wise' in (1) or, in Frege's examples, when we substitute 'Alexander' for 'a mammal', or 'Venus' for 'green', we simultaneously but covertly change the use of 'is'. Instead of being the copula, it now means 'is identical to', which is the copula followed by an unsaturated dyadic predicate (1892a: 44, and first footnote).

The conventionalist account permits singular terms, without change of meaning, to appear in atomic sentences as one-place predicates. For example, (3) is an atomic sentence with 'Socrates' as subject in its first occurrence and as predicate in its second occurrence, in which being Socrates (or Socrateity) is predicated of Socrates. This holds not only for concrete singular terms such as 'Socrates', but also for abstract singular terms. For example, in contrast to the general term 'wise', which connotes wisdom and denotes wise things, the singular term 'wisdom' connotes being wisdom and denotes wisdom. The sentence 'Wisdom is wisdom' is true if and only if the denotation of the subject, wisdom, exemplifies the connotation of the predicate, namely, being wisdom. This account is therefore not committed to there being an ambiguity in 'is' as between predication and identity.

It has been argued (Frederick 2013) that Frege's prohibition on singular terms playing the role of predicate impairs the power of his logic and that 'is' is not in fact ambiguous between predication and identity. Here we need only note the different treatment of singular terms in the two theories, as it is relevant for the discussion in sub-section 4.3.

4.3 Paradox

On Frege's view (1891, 1892a, 1892b), an expression's sense is a way of determining the expression's reference. An expression's reference is what it stands for. The reference of a sentence is its truth-value. The reference of a singular term Frege calls an 'object'; the reference of a predicate he calls a 'concept'. The reference of a complex expression is determined by the references of its parts. Therefore, if a singular term had the same reference as a predicate, then the former could be substituted for the latter in a sentence without changing the truth-value of the sentence. However, as we noted in sub-section 4.2, on Frege's view, substituting a singular term, which has a complete sense, for a predicate, which has an unsaturated sense, turns a sentence into a list, and thus does not preserve its truth-value. Therefore, a singular term and a predicate cannot have the same reference: no object is a concept. Frege makes this point by saying that concepts are unsaturated or incomplete, while objects are complete. However, this leads to paradox because we can construct definite descriptions which ought to denote concepts but which, since they are singular terms, can denote only objects. As Frege says: 'The three words "the concept 'horse'" do designate an object, but on that very account they do not designate a concept' (1892a: 45); 'the concept *horse* is not a concept' (1892a: 46). Some theorists, following Russell (1905), do not count definite descriptions as genuine singular terms; but an appeal to this would not save Frege from paradox because we can introduce proper names for the references of predicates. Although Frege discusses the paradox only in connection with the references of predicates, a counterpart paradox arises in connection with their senses. For, as an unsaturated sense is an incomplete entity, it cannot be referred to by a singular term, which can denote only a complete entity. Thus, the sense of the predicate 'wise' (or 'is wise') is not a sense of a predicate.

Frege proposed to eliminate the paradox by proscribing singular terms referring to concepts, so that whenever we want to say something about a concept, we would have to find other ways of saying it (1892a: 48–50; 1892–95: 122). For example, if we want to say that the concept *horse* has some instances, we could say 'there are horses'. The proscription could be extended to singular terms referring to unsaturated senses. But all that is entirely ad hoc: the only reason for adopting the proscription is to save the theory.

The conventionalist account avoids Frege's paradoxes, since it posits no such things as unsaturated senses or references which cannot be denoted by singular terms. Terms connote properties and denote the things that exemplify those properties. Neither properties nor the things that exemplify them are unsaturated; either may be denoted by singular terms. As we saw in sub-section 4.2, the singular term 'wisdom' denotes the property wisdom. The predicative sign (the copula or the form of the verb, in English and in many other languages) is not a term: it connotes no property, so it can denote nothing. Its meaning is syncategorematic: it is explained in terms of the truth-conditions of the sentences in which may occur. However, the fact that it is not a term does not imply that its meaning is unsaturated and therefore incapable of being denoted by a singular term. The meaning of the predicative sign is denoted by the singular term,

(p) the meaning of the predicative sign

which connotes *being the meaning of the predicative sign*. If we like, we can even introduce a proper name for the meaning of the predicative sign. Once we recognise syncategorematic expressions which are given functions by linguistic conventions, we solve the problem of the unity of the sentence without the use of the obscure and mystical notion of unsaturatedness. There is no hint of paradox here.

Peter Geach (1950: 476) argues that the claim that singular terms may be predicates generates a version of Russell's Paradox. The relevant aspect of Geach's argument can be stated as follows. If we allow sentences of the form 'A is A', where 'A' represents a singular term, then we must admit sentences of the form 'A is not A' as well-formed (though false); for example, 'Socrates is not Socrates'. Consider now an open sentence, 'x is not x'. That expresses a property of the object indicated by the variable 'x'. If we introduce a term, 'W', to express that property, then we have:

(‡) for any x, x is W if and only if x is not x.

However, since the same term may be either predicate or subject, we may substitute 'W' for the variable 'x' in (‡) to obtain the contradiction:

(*) *W* is *W* if and only if *W* is not *W*.

Recall from section 3 that, on the conventionalist account, while singular terms may be subjects or predicates, it is only singular terms that may be subjects. So, if substitution in (‡) is to yield (*), '*W*' must be a singular term. Geach's talk of '*W*' *expressing* a property is unclear. The property in question is the-property-of-being-not-identical-to-itself. On the conventionalist account, the term '*W*' is introduced either to connote or to denote that property. Let us suppose, first, that '*W*' *connotes* the property. It then *denotes* each thing which is not self-identical, that is, it denotes nothing. On that construal, (‡) is true: anything exemplifies the-property-of-being-not-identical-to-itself if and only if it is not itself. However, with that meaning, '*W*' is one of those general terms which cannot denote anything, rather than a singular term, so it cannot play the role of subject, in which case (*) is not well-formed, so Geach is mistaken in saying that (*) follows from (‡).

Let us suppose, second, that '*W*' *denotes*, rather than connotes, the-property-of-being-not-identical-to-itself (and thus connotes *being the-property-of-being-not-identical-to-itself*). In that case, (‡) is false because there is something, namely, the-property-of-being-not-identical-to-itself, of which the left-hand-side of the bi-conditional is true and the right-hand-side of the bi-conditional is false. That is, it is true that the-property-of-being-not-identical-to-itself is the-property-of-being-not-identical-to-itself; but it is false that the-property-of-being-not-identical-to-itself is not the-property-of-being-not-identical-to-itself. So Geach is mistaken in maintaining that (‡) follows from the combined assumptions that singular terms may be predicates and that a singular term may denote the-property-of-being-not-identical-to-itself. He has therefore not shown that (*) follows from those assumptions.

Thus, Geach's argument fails: introducing a term to connote or to denote the-property-of-being-not-identical-to-itself does not generate a contradiction in the conventionalist account. Of course, this is not to say that the conventionalist account resolves Russell's Paradox. That paradox will still arise in the familiar ways unless steps are taken to prevent it, whether we adopt the conventionalist account or Frege's unsaturatedness account (Russell 1959: 58–59). But Geach has not shown that the conventionalist account generates a paradox of its own, so the conventionalist account still has the advantage over Frege's in not being afflicted by Frege's paradoxes of sense and reference.

5. Faltering Steps

Some recent philosophers have taken more or less faltering steps in the direction of the conventionalist account of the unity of the sentence; but they have been hampered by their adherence to a predominantly Fregean framework of thought. It may help to clarify the conventionalist account if I contrast it with the accounts given by three of these philosophers, though I can do so only briefly here.

Peter Strawson (1974: 20–22) offers a conventionalist account of sentential unity along the lines of Mill, but with some significant differences. He distinguishes paradigm singular terms as those expressions which stand for spatio-temporal particulars and paradigm predicates as those which stand for properties of such particulars and he then recognises various types of non-paradigm singular terms and predicates on an analogical basis. However, he does not distinguish the connotation and the denotation of terms. As a consequence, he cannot recognise that singular terms may appear as predicates, he can accommodate non-denoting singular terms only in an ad hoc way, he falls into inconsistency, and he commits to a cumbersomely stratified type-hierarchy (see Frederick 2011).

David Wiggins (1984: 133) suggests that the copula should be separated from the general term into which Frege sometimes absorbs it and treated syncategorematically. Using 'term' to mean 'singular term', he offers a convention similar to Mill's:

> Where t is a term and PRED is any predicate expression of the form of VERB, or ADJECTIVE, or 'a' + SUBSTANTIVE, True [t + copula + PRED] iff PRED is true of the designation of t (or, as Frege might have preferred to say, the designation of t falls under the concept that PRED stands for).

This convention is significantly different from convention (C1), in section 3, above. First, it says nothing of relational terms. Second, it does not permit singular terms to appear as predicates. Third, it concerns only a fragment of English, rather than being a general account of the meaning of the sign of predicative combination in any language that has atomic sentences. Fourth, Wiggins says that the convention explains the meaning of the copula, not by itself, but only in conjunction with the conventions assigning meaning to each primitive verb, adjective and substantive of the fragment of English concerned. This has the peculiar and unwelcome consequence that if we add a new primitive term or change the meaning of any of the existing primitive terms (as happens in the development of scientific theories, for example), we thereby change the meaning of the copula. However, the more baffling parts of Wiggins' account are yet to come.

Wiggins recognises that the designation of a complete entity followed by the designation of an incomplete entity would be a list, not a sentence (1984: 139). Accordingly, he proposes (1984: 133-134) that:

- a predicate has a 'semantic value' which is the concept that it stands for, but the concept is not unsaturated;
- an expression formed by combining a predicate with the copula, such as 'is wise' or 'is a man', has a semantic value which it does *not* stand for, namely, a function-in-extension from objects to truth-values, which is also not unsaturated.

However, he says, an expression of the latter type is "an unsaturated expression that will in its turn combine in the fashion Frege himself describes with a saturated expression to produce a complete sentence" (1984: 133). It must be the *sense* of such an expression that is supposed to be unsaturated, since the expression has no reference and its semantic value is not unsaturated, and since Wiggins seems to accept that simple, as opposed to complex, expressions of the form 'is F' are not syntactically incomplete (1984: 141, n18). Three questions cry out for an answer. First, why not say simply that the meaning of such an expression is derived from the conventions, mentioned in the previous paragraph, that assign the truth-condition of monadic atomic sentences and the meanings of the predicates? In other words, why do we need the function-in-extension to be its 'semantic value' (and what does it even mean to say that)? Second, and more important for our purpose, our convention (C1) solves the problem of the unity of the sentence, and Wiggins' counterpart conventions would do the same if they worked, so why does he retain the unnecessary mystical notion of unsaturatedness? Third, given that he sees that unsaturated references would be useless in solving the problem of the unity of the sentence, why does he not see that unsaturated senses would also be useless, for the same reason? The deficiencies of Wiggins' obscure account seem to be the result of attempting to repair Frege's theory instead of trying to replace it.

Donald Davidson rejects Frege's unsaturatedness theory (2005: 156), but his proposed solution to the problem of the unity of the sentence appears to consist of a collection of more or less relevant reflections with no clear point. He says that predication, and thus the unity of atomic sentences, relates to truth and to logical form; that the copula does not stand for a relation; that predicates do not stand for abstract entities but are true of, or false of, things; and that Tarski's truth-definition for a (simple formalised) language can be adapted to state the truth-conditions of the sentences of such a language (2005: 141-154). With the exception of his negative claim about abstract entities, few logical theorists would deny those things or would need to be reminded of them. So how are they supposed to provide a solution to the problem of the unity of the sentence? It seems that what Davidson takes to be his key insight is the following.

> The importance of the connection [between truth and predication] is this: if we can show that our account of the role of predicates is part of the explanation of the fact that sentences containing a given predicate are true or false, then we have incorporated our account of predicates into an explanation of the most obvious sense in which sentences are unified, and so we can understand how, by using a sentence, we can make assertions and perform other speech acts (Davidson 2005: 155).

That is a convoluted sentence. So far as I can make out, all it says is that, if we can show how predicates contribute to the truth-conditions of sentences, we will have solved the problem of the unity of the sentence. That much, as we saw in section 3, had been said, and said much more clearly, by Mill and by generations of logicians who preceded him. Of course, Davidson is impressed by the fact that Tarski's truth-definition can be adapted to generate the truth-conditions for each sentence of a language (2005: 154–160). But it can do that only for artificial, strictly regimented languages. It is difficult to see what Davidson has to teach us about the problem of the unity of the sentence.

6. The Unity of the Proposition

The conventionalist theory, being about linguistic conventions, offers no solution to the problem of the unity of the proposition. However, following a suggestion of Manuel García-Carpintero (2010: 288), I hypothesise that that problem has no solution. The problem of the unity of the proposition arises from a supposition that a proposition can be analysed into non-propositional constituents. The supposition is that it should be possible to give a reductive analysis of propositions into things which are not propositions, an analysis which would enable propositions to be explained away, or dispensed with in favour of more familiar entities. Yet it is difficult to see how such a reductive analysis is possible, given that a proposition, unlike its supposed constituents, either by themselves or as a collection, *says something* which is capable of being true or false.

In consequence, propositions should be accepted as a distinct class of entities, expressed by sentences, which exist over and above the properties which are connoted by terms. A proposition is *related* to properties, in the way that the proposition expressed by (1) is true if and only if the property of wisdom is exemplified by the thing which exemplifies the property of Socrateity; and we *identify* propositions by means of the properties to which they are related. But a proposition is not *made up* of properties, or of anything else. Thus, the problem of showing how properties can be combined to produce a proposition that is true or false is a spurious problem: propositions cannot be produced from such combinations. The reason a proposition disappears when supposedly analysed into its constituent entities is that it has no constituent entities: the entities paraded as constituents are in fact separate entities which are related to the proposition but are not constituents of it. A so-called analysis of a proposition is therefore not an identification of its components.

It might be urged that this attempted dissolution of the problem faces a counterpart problem of an essentially similar kind. While we no longer have the problem of explaining how the elements of a proposition can combine to form a proposition, we now have to explain how properties can be related in the way required for there to be a proposition. For instance, what accounts for the ability of the properties, Socrateity and wisdom, to be related in such a way that the (non-composite) proposition that Socrates is wise is true if and only if the thing which exemplifies Socrateity also exemplifies wisdom?

The answer is that the existence of the proposition accounts for it, because the nature of that proposition is such that it is true if and only if the properties are related in that way. To assume that the question can be answered without reference to the proposition is mistakenly to assume that propositions can be reduced to non-propositional entities. The 'ability' of properties to be related in the appropriate way demands nothing of the properties themselves, except that they exist. We can see this if we consider an impossible proposition, say, that the number two is odd. That proposition is related to the properties of two-ness and oddness in such a way that it is true if and only if the thing that exemplifies two-ness also exemplifies oddness. The proposition is related to the properties in that way even though it is impossible that the thing which exemplifies two-ness also exemplifies oddness.

It is, therefore, a separate question what accounts for the *truth* of a true proposition, say, that Socrates is wise. An uninformative answer to that question is that the thing which exemplifies Socrateity exemplifies wisdom; or, more prosaically, that Socrates is wise. A more informative answer would be a set of propositions which together explain why Socrates is wise. However, we cannot say what makes a proposition true without stating a proposition. Propositions are not reducible to properties or relations between properties, because they say something.

We must therefore admit distinct levels of the abstract realm: there are properties, which may be exemplified by things but cannot be true or false; and there are propositions, which may be true or false. There are also arguments, which may be valid or invalid, and which are not composed of propositions, though they are related to propositions; but that is another topic.

7. Conclusion

Frege attempts to solve the problems of the unity of the sentence and the unity of the proposition by postulating a property of unsaturatedness. The attempt fails because:

- it is ad hoc, being simply a mystical re-description of the problems;
- it does not even work, because the problems recur even if the theory of unsaturatedness is accepted;
- it is inconsistent with the fact that we can list expressions which, on Frege's account, stand for complete and unsaturated entities;
- it debars singular terms from appearing as predicates and requires a spurious ambiguity in 'is';
- it generates its own paradoxes which can be avoided only by ad hoc manoeuvres.

In consequence, although Frege is arguably the greatest logician who has lived to date, and his symbolic logic is a major advance over the systems that went before it, his doctrine of unsaturatedness is a piece of pseudo-scientific occultism that has no place in the serious development of logical theory.

The problem of the unity of the proposition appears to be a pseudo-problem. It seeks an explanation for how non-propositional constituents combine to form a proposition; but propositions seem to be entities that have no constituents.

The traditional logicians solved the problem of the unity of the sentence by noting that some linguistic expressions, or forms of expression, are conventionally given the function of signifying predicative combination, unlike other expressions which are given the functions of singular or general terms. The conventionalist account given in section 3 improves on the traditional theory by accommodating three Fregean insights, namely, that atomic sentences may be relational, that complex terms are derived from sentences, and that predications may be formulated without being asserted or denied. It also assigns connotation to names, as had been done by some traditional logicians, thereby streamlining Mill's version of the traditional account. This conventionalist account is a simpler account than Frege's; it can be stated in plain and intelligible language involving no mystical posits; it has genuine explanatory value; it is free of Frege's paradoxes; and it permits singular terms to play the role of predicates and thereby avoids positing a special 'is' of identity.

Logical theorists who have been brought up in the Fregean tradition (almost certainly, everyone who reads this paper) might wonder whether the conventionalist account is itself committed to ascribing unsaturated senses to terms. For, following Frege's lead, the account distinguishes terms as monadic, dyadic, triadic, and so on, according to the number of singular terms with which they must be combined to produce a sentence. But does not this n-adic nature of terms signify their incompleteness and give them an unsaturated sense? How else can the n-adic nature of terms be explained?

First, the explanation for the n-adic nature of terms seems straightforward. Things have (monadic) properties and (polyadic) relations, and some relations hold between two things, some hold between three, and so on. So if we are to talk about things, their properties and the relations that they stand in, we need terms that connote the various properties and relations. We have conventions assigning such connotations to simple terms and we can derive the connotation of complex terms from the sentences in which they occur. Second, the n-adic nature of terms clearly does not give them an unsaturated sense if 'unsaturated' is meant in Frege's sense. We recognise no terms which have senses such that the terms will combine with some other terms to form atomic sentences independently of a linguistic convention specifying when such a predicative combination occurs; and we recognise no senses or other items which cannot be denoted by singular terms. Indeed, we recognise that *singular terms* are monadic terms which may be combined with singular terms to form sentences. Of course, all terms, including singular terms, are incomplete in the sense that they are not by themselves sentences: to get an atomic sentence from terms we need a sign indicating that the terms make a composite which is true under a specified condition. The Fregean may remonstrate: does not *this* give *a* sense in which the senses of terms are unsaturated? If we were committed to using the word 'unsaturated,' that would indeed give an intelligible sense to it; but it would be confusing to use the word 'unsaturated' for that purpose because it already has an established sense (Frege's) which is not only different but is also a barrier to clear thinking. It would be like retaining the term 'dephlogisticated air' as a label for oxygen.

References

Arnauld, A. and Nicole, P. (1662). *Logic or the Art of Thinking*. Tr. J. Dickoff and P. James. Indianapolis: Bobbs-Merrill (1964).

Davidson, D. (2005). *Truth and Predication*. Cambridge, MA: Belknap Press.

Dummett, M. (1981a). *Frege: Philosophy of Language* (2nd ed.). London: Duckworth.

Dummett, M. (1981b). *The Interpretation of Frege's Philosophy*. London: Duckworth.

Frederick, D. (2011). "P. F. Strawson on Predication". *Polish Journal of Philosophy* 5 (1): 39–57.

Frederick, D. (2013). "Singular Terms, Predicates and the Spurious 'Is' of Identity". *Dialectica* 67 (3): 325–343.

Frege, G. (1891). "Function and Concept". In: P. Geach and M. Black (eds.), 1980, 21–41.

Frege, G. (1892a). "On Concept and Object". In: P. Geach and M. Black (eds.), 1980, 42–55.

Frege, G. (1892b). "On Sense and Meaning". In: P. Geach and M. Black (eds.), 1980, 56–78.

Frege, G. (1892–95). "Comments on Sense and Meaning". In: G. Frege, 1979, 118–125.

Frege, G. (1897). "Logic". In: G. Frege, 1979, 126–51.

Frege, G. (1903). "On the Foundations of Geometry". Tr. M. E. Szabo. In: E. Klemke (ed.), *Essays on Frege*. Urbana: University of Illinois Press (1968), 559–575.

Frege, G. (1906a). "Plan of a Critique of Schoenflies". In: G. Frege, 1979, 176–183.

Frege, G. (1906b). "Introduction to Logic". In: G. Frege, 1979, 185–196.

Frege, G. (1914). "Logic in Mathematics". In: G. Frege, 1979, 203–250.

Frege, G. (1915). "My Basic Logical Insights". In: G. Frege, 1979, 251–252.

Frege, G. (1979). *Posthumous Writings*. H. Hermes, F. Kambartel and F. Kaulbach (eds.), tr. P. Long and R. White. Oxford: Blackwell.

García-Carpintero, M. (2010). "Gaskin's Ideal Unity". *Dialectica* 64 (2): 279–88.

Geach, P. (1950). "Subject and Predicate". *Mind* 59 (236): 461–82.

Geach, P. and B, Max. (eds. and tr.). (1980). *Translations from the Philosophical Writings of Gottlob Frege*, third edition. Oxford: Blackwell.

MacBride, F. (2005). "The Particular-Universal Distinction: A Dogma of Metaphysics?" *Mind* 114: 565–614.

MacFarlane, J. (2009). "Logical Constants". In E. N. Zalta (ed.), *The Stanford Encyclopedia of Philosophy (Fall 2009 Edition)*. Downloaded 2 May 2010 from: http://plato.stanford.edu/archives/fall2009/entries/logical-constants/.

Mill, J. St. (1843). *A System of Logic*. J. M. Robson (ed.), Toronto: University of Toronto Press (1974). Downloaded 2 May 2010 from: http://files.libertyfund.org/files/246/Mill_0223-07_EBk_v5.pdf.

Roux, S. (2013). "An Empire Divided: French Natural Philosophy (1670–1690)". In: D. Garber and S. Roux (eds.), *The Mechanization of Natural Philosophy*. London: Springer, 55–95.

Russell, B. (1905). "On Denoting". In: R. C. Marsh (ed.), *Logic and Knowledge*. London: George Allen and Unwin (1956), 41–56.

Russell, B. (1959). *My Philosophical Development*. London: George Allen & Unwin.

Strawson, P. (1974). *Subject and Predicate in Logic and Grammar*. London: Methuen.

Wiggins, D. (1984). "The Sense and Reference of Predicates: a Running Repair to Frege's Doctrine and a Plea for the Copula". In: C. Wright (ed.), *Frege: Tradition and Influence*. Oxford: Blackwell, 126–143.

David Liebesman
University of Calgary

Sodium-Free Semantics: The Continuing Relevance of The Concept *Horse*

Abstract: Far from being of mere historical interest, concept horse-style expressibility problems arise for versions of type-theoretic semantics in the tradition of Montague. Grappling with expressibility problems yields lessons about the philosophical interpretation and empirical limits of such type-theories.

Keywords: Gottlob Frege, concept *horse*, type-theory, semantics

1. Concept *Horse* Problems in Frege

Famously, Frege hit a wall when attempting to discuss predicate reference.[1] His exclusive distinction between concepts and objects, when conjoined with the thesis that the former are uniquely suitable for predicate reference and the latter are uniquely suitable for non-predicate reference, leads to familiar problems. Assume that we want to discuss the referent of a predicate, e.g. *horse*. In order to do this, it is natural to attempt to introduce a non-predicate that refers the predicate's referent, e.g. *the concept horse*. Unfortunately, Frege's semantic theses undermine such a move. If *the concept horse* is a non-predicate then it can't, by Frege's lights, co-refer with a predicate. It seems, then, that we are left without the terms we need in order to adequately discuss the semantics of predicates.

This problem has engendered the full range of reaction. Some think that Frege's concept/object distinction is misguided and the resultant problems are avoidable.[2] Others think that the distinction reflects the true nature of logical categories and therefore some such problems are unavoidable.[3] I will argue that versions of the problem arise for type-theoretic semantics in the tradition of Montague and that grappling with the problems yields insight into the empirical limits and philosophical understanding of such theories. In the remainder of this section I'll discuss concept horse problems in Frege. In section 2, I'll argue that some

---

1   The most famous and relevant passage is his (1892: 182–185).
2   E.g. Burge (2005: 19–21).
3   E.g. Burgess (2005), Proops (2013), and Trueman (2015).

analogous problems arise for a simple Montague-inspired type-theory. In section 3, I'll consider various ways to complicate the theory and show how these don't obviously solve the problems. My main contention is simply that the problems remain problems for type-theoretic semantics. My secondary contention, developed in section 4, is that the most promising solution distinguishes between semantic values and ways of having those semantic values.[4] This allows expressions with the same semantic value to be semantically differentiated by the ways they have those values.

1.1 Three Problems

Following Hale and Wright (2012), we can see concept horse problems as arising from a principle that links ontological categories with linguistic categories, along with another principle about the nature of semantic categories.

> **Correlation**: an entity can, and can only, be referred to with an expression of a correlative semantic type.[5]
>
> **Category**: there is a way of dividing words/occurrences of words into mutually exclusive categories such that Correlation is true of that categorization. Furthermore, this categorization plays an important philosophical/explanatory role.

Correlation means little until we understand the notion of semantic types that it utilizes. There are myriad ways to divide words into semantic types. For instance, we could divide words into types on basis of whether they designate things I like. In this type-system, *dogs* and *cats* would be distinct types. Such a categorization, for Frege, is not a plausible basis for deriving ontological conclusions. Rather, we need a categorization of words (occurrences) on basis of predicate/argument structure, i.e. what is a predicate and what is an argument. One reason that Frege takes argument structure to be important is that he takes it to hold the key to understanding the structure of thoughts:

---

4   My favored version of this view is developed in Liebesman (2015). Other versions of the view are considered in Furth (1968), Strawson (1974), Wright (1998), Burge (2005), Hale and Wright (2012), MacBride (2011), and Rieppel (forthcoming).
5   **Correlation** differs slightly from the principle on which Hale and Wright focus. Hale and Wright swap *semantic* for *logico-syntactic* in their articulation of the principle. For their purposes, and perhaps for Frege's, we can make inferences directly from logical category and syntactic category to ontological category. By contrast, I will remain agnostic about whether there are syntactic categories/distinctions which are semantically inert.

For not all parts of a thought can be complete; at least one must be unsaturated or predicative; otherwise they would not hold together (Frege 1997: 193).

Category helps us isolate the type-system relevant to **Correlation**. In fact, one may think of **Correlation** as a schema, with instances provided by different type systems. **Category** allows us to focus on the relevant instance: that in which types are determined by argument structure. I will further discuss what this means in section 3, and I discuss it in more detail in Liebesman (2015). For the remainder of the discussion, I'll use *Correlation* to pick out the principle on which types are determined by argument-structure.

Frege's thoughts are composed of senses, though he also held that the typing of senses is mirrored both by referents, and by words themselves. Our semantic typing will primary concern the level of reference. In section 3.5 I'll discuss how Frege's concerns about the structure of thoughts can be generalized beyond sense.

The combination of **Correlation** and **Category** entails that, given some words/occurrences and the entities to which they refer, we can divide the entities into mutually exclusive ontological classes on basis of the division of those words/occurrences into mutually exclusive semantic classes (of the relevant sort). This leads to at least three problems.[6]

First, and most familiar, is the breach of custom problem(**BOC**).

BOC: Frege's semantics commits him to endorsing such bizarre sentences as *The concept horse is not a concept.*

If we stipulate that *concept* picks out whatever can serve as the referent of a predicate, then we can derive the problem as follows:

1. *The concept horse* is not a predicate. (assumption about the semantics of definite descriptions)
2. Predicates refer to concepts. (from the definition of *concept*)
3. The referent of *the concept horse* is not a concept. (1, 2, and Correlation)
4. Therefore, the concept horse is not a concept. (4, disquotation)

There are numerous ways one may wish to solve **BOC**. Many of these will be discussed in detail in sections 2 and 3. However, I will not focus on **BOC** itself any further. The reason is that some obvious ways to solve **BOC** will not solve other closely related problems. In particular, one may wish to solve **BOC** by rejecting premise 1 and taking definite descriptions to be predicates, or by rejecting the

---

6 My taxonomy is inspired by Proops (2013), though it differs as it is tailored for my purposes.

disquotational principle that allows us to infer 5 from 4. Whether or not these strategies are promising for dealing with **BOC**, they do nothing to resolve the deeper underlying problems.

The other two problems I will consider are broader that **BOC**.[7] These problems purport to show that a large class of sentences cannot be adequately analyzed given Frege's semantic commitments. The second problem is metasemantic: the class of sentences that cannot be adequately analyzed are sentences about semantics. The third problem pertains to any sentence that requires upholding intimate connections between predicates and non-predicates. Some such sentences may be about semantics (e.g. denotation specifications) but, as we'll see, there are many such sentences that aren't themselves about semantics.

We'll label the second problem Metasemantic Failure (**MF**):

> **MF**: Frege's semantic theses do not allow him to adequately analyze sentences about semantics.

Before showing how Frege falls prey to **MF**, we need to clarify the problem. If we take semantic analyses to match sentences with truth-conditions (perhaps by way of translating them into another language, perhaps not), then a necessary condition on adequate analysis is truth-conditional adequacy. As we'll see, the problem for Frege is that, by his own lights, his analyses fail to meet this condition. There are surely other necessary conditions on adequate analysis –

I'll discuss one in the next section – but, for our purposes in this section, truth-conditional adequacy will suffice.

We've clarified the adequate analysis component of **MF**, what about the target of those analyses – sentences about semantics? The form of a Fregean semantic theory is familiar: we assign semantic values from a type-theoretic hierarchy to each meaningful expression and combine them using function-application. So, at the very least, a theory will be constituted by sentences that ascribe semantic values. Just which sentences are covered will depend on which relation is involved

---

7   This overview of the concept *horse* problem in Frege is both brief and biased. It is brief because my main aim is not to consider the problem in Frege, it is to argue that analogous problems arise for contemporary theorists. It is biased because in discussing Frege I focus on problems which have contemporary analogs. There are lots of other problems for Frege in the vicinity that I do not discuss. For instance, one may hold not merely that Frege couldn't adequately analyze certain metalinguistic sentences, but that he was left without the resources to express metalinguistic facts altogether. Frege may well run into this problem, but I don't see how it would straightforwardly generalize so I set it aside.

in ascription of semantic value. Some familiar candidates are reference, designation, and interpretation. In eliciting the problems, I'll begin with reference. In 3.1 I'll consider whether distinguishing these relations can help to solve the problem.

Given this, the analysandum relevant to **MF** will include sentences of the form *w refers to v* where *w* picks out a word and *v* its referent. However, **MF** shouldn't be understood as limited to such straightforward reference-specification. Metalinguistic discourse is constituted by sentences beyond those included in a semantic theory. In the course of theorizing, for instance, we will deny that words have certain values, compare the values of words, and circumscribe our overall theory by claiming that it captures all meaning-facts.

The problem is that Frege's semantic theory appears to prevent us from adequately articulating myriad reference-specifications, in addition to exhaustivity clauses, comparisons, and denials. For instance, assume we wish to correlate a predicate P with its referent. The natural way to do this is to introduce a singular term that co-refers with P. That way we can straightforwardly claim, for instance, that *horse* refers to the concept horse. Frege's views prohibit this. *The concept horse* is a non-predicate, so by **Correlation** and **Category**, it cannot co-refer with *horse*, which is a predicate.

To combat this problem, some theorists attempt to identify other constructions better suited for articulating semantic theories. I am skeptical here, for reasons familiar in the literature.[8] However, even if we identify other constructions better suited for specifying reference, there are principled reasons to doubt that those constructions would be suited for articulating denials and comparisons. We can see this by getting more precise about the distinctive philosophical/explanatory role mentioned in Category.

However we understand the problem of the unity of the proposition it is clear that, for Frege, solving the problem is one main philosophical/explanatory roles played by the relevant division of words into semantic types and their referents into ontic types. Let's make this more explicit with Explanation.

Explanation: the division of words into mutually exclusive semantic types that entails correlative ontic types explains how words can combine to express thoughts. In particular, a word with a referent of a particular type can only combine with words with referents of two relevant types: those suitable to be its arguments and those suitable to take it as an argument.

---

8 Dummett (1973) is the most discussed proposal. See Wright and Hale (2012) for a litany of objections.

Explanation, which summarizes Frege's method of typing words and referents, places restrictions on which sentences express thoughts, i.e. are interpretable. These restrictions entail that we cannot articulate the denials and comparisons we need in order to conduct semantic discourse.

Begin with denials. For example, we wish claim that (1) and (2) are false, and that (3) and (4) are true.

(1) The denotation of *wise* is an object.
(2) The denotation of *wise* is not of predicate type.
(3) It is false that the denotation of *wise* is an object.
(4) It is false that the denotation of *wise* is not of predicate type.

Correlation, Category, and Explanation prevent us from adequately analyzing (1)–(4). The reason is that the predicates *is an object* and *is not of predicate type* require non-predicates as their arguments. This is made clear by (5) and (6)

(5) Socrates is an object.
(6) Socrates is not of predicate type.

This leaves only three options in analyzing (1)–(2). The first option is that the sentence-initial definite descriptions in those sentences designate objects. The problem with this option is that the sentences end up being true. So, the option fails to be truth-conditionally adequate. The second option is that the sentence-initial definite descriptions designate concepts. In this case, the sentences are ill-formed, and therefore don't express thoughts at all. (One may invoke type-shifting to respond to this worry; I will discuss type-shifting in detail in section 3.) Thus, they fail to be either true or false, again failing truth-conditional adequacy. The third option is that *is an object* and *is not of predicate type* do not refer to the same concepts in (1)/(2) as they do in (5)/(6). We cannot immediately fault this proposal on truth-conditional adequacy alone, given that the third view does not yet contain an account of what the predicates designate. However, such a view will fail to provide truth-conditionally adequate analyses for comparisons, such as (7) and (8) in which we use a single predicate to express the distinctions between predicates and non-predicates.

(7) The denotation of *Frege* is an object, while the denotation of *wise* is not.
(8) The denotation of *Frege* is not of predicate type, while the denotation of *wise* is.

(7) and (8), given their explicitly comparative nature, foreclose the possibility of predicate ambiguity. Thus, we are left with only the first two options, both of which fail truth-conditionally. (One could attempt some fancy semantic footwork

and claim that the elided predicates in (7) and (8) receive different readings akin to sloppy interpretations of anaphoric pronouns. However, there is no precedent for changing the semantic type of the elided expression, which is what this would require.)

The reason that these problems arise is that when articulating a theory about predicate reference we do not wish to limit ourselves to mere reference-specification. We also wish to say what particular predicates do not refer to, as well as how they compare. Even at their most successful, Fregean attempts to articulate semantic theses about predicates will fail here. Such attempts consist in identification of special, peculiar expressions, which allow us to formulate those claims. Limitation to such peculiar expressions, by its nature, rules out any attempt to use ordinary expressions to express falsehoods, or true negations.

We'll label the third problem Nominalization Failure (NF).

NF: Frege's semantic these prevent him from adequately analyzing sentences containing predicate nominalizations.

While MF focuses squarely on discourse about semantics, NF is broader. There are several natural language constructions that, intuitively, allow us to use singular terms to refer to predicate referents, for garden-variety non-theoretical discourse. Consider the following sentences.

(9) Swimming is a fun activity.
(10) To think is to be human.
(11) Wisdom is a feature of my favorite philosophers and linguists.

None of (9)-(11) is a plausibly candidate for inclusion in a semantic theory. Nonetheless, they all look troublesome given Correlation and Category. The trouble stems from the fact that the most straightforward account takes their subject terms to refer to entities that, in other contexts, are predicate referents. Even if a theorist did the implausible and found a idioms sufficient for articulating a semantic theory in a way consistent with **Correlation** and **Category**, they may not have succeeded in giving a plausible account of the full range of discourse in which we seem to use non-predicates to refer to predicate referents.

For the Fregean to meet this challenge, they would have to give a semantics for predicate nominalizations that is consistent with Category and Correlation. Thus far, I've only mentioned a *prima facie* reason to think this can't be done. I won't discuss possible solutions for Frege in any great detail, though I will examine this problem in more detail when it discuss it in the context of contemporary type-theory.⁹

Stepping back, we can see that there are a range of potential solutions to MF and NF. A modest solution would be to find articulate reference-specifications. A less modest solution would find a way to analyze all sorts of metalinguistic sentences, which will plausibly include generalizations, denials, as well as claims about predicates beyond merely what they refer to. A comprehensive solution will find a way to adequately analyze the full range of relevant discourse, this includes understanding claims like (9)-(11).

Seen in this light, it is reasonable to desire a comprehensive solution. After all, what's the alternative? Denying the intelligibility of a wide swath of common sentences that, axes aside, appear perfectly intelligible, seems untenable. A comprehensive discussion of these problems would require surveying the potential Fregean solutions. There's a substantial literature on just this. My aim, however, is different. I will not assess the severity of the problems for Frege. Rather, I'll turn my attention to contemporary type-theoretic semantics.

2. Concept *Horse* Problems in Type-Theory

My next goal is to argue that versions of **MF** and **NF** arise in contemporary type-theoretic semantics. The goal raises a worry. Type-driven semantic theories can be articulated in varied ways, with varied resources, and varied commitments. Any attempt to boil such a rich and diverse research project into a few pithy claims runs the risk of ignoring important theoretical distinctions and complications. To combat this, I'll begin with a deliberately simple type-theory of the sort familiar from introductory textbooks.¹⁰ I'll argue that versions of **MF** and **NF** arise for this view. I'll then consider various ways to complicate the simple theory and whether they help solve the problems.

In our simple theory, there are two basic types: <e> and <t>. Complex types are derived by ordered-pair formation: any ordered-pair of types is itself a type. More complicated type-theories add basic types for additional categories. Without delving into complications, e.g. Montague's treatment of all noun phrases as having type-identical meanings, it is natural to assign meanings of type <e> to referential terms, and type <t> to truth-apt sentences. More complex types of meanings are assigned to predicates that map entities to truth-values, and combine

---

9   Liebesman (2015) contains a more detailed critical discussion of Fregean attempts to analyze predicate nominalization.
10 Heim and Kratzer (1998) is the most familiar example.

with non-predicates to form sentences. For example, *is swimming* is assigned a meaning of type <e,t>: those who are swimming are mapped to truth and all others are mapped to falsity.[11] Every simple lexical item is assigned a meaning that falls in this type-theoretic hierarchy. To calculate the meanings of complexes, our type-theory contains a single rule of composition: function-application. Taking double-bracket notation to signify semantic value, our general rule of composition is [[AB]] = [[A]][[B]], for any expressions A and B. Particular instances of function-application are type-driven; i.e. which is the function and which is the argument is determined by the types of the meanings being composed, rather than the order in which they appear.

We can consider types of meanings as well as types of linguistic items. The function λx.x is swimming, which is type <e,t>, is the meaning of *is swimming*. It is also common to claim that *is swimming* – the linguistic item rather than its semantic value – has type <e,t>. Note, though, that moving freely between the types of meanings and the types of linguistic items requires a correlation between them. Since the viability of such a correlation is one of the issues being considered, I will reserve types such as <e,t> for meanings and use *non-predicate, predicate, first-order predicate*, etc. for classifying linguistic types.

Our simple theory adopts two assumptions about the type system:

Exclusivity: no meaning belongs to more than one type.
Inflexibility: each word is assigned a single meaning (from a single type) and this is its only semantic contribution.

Later I will examine these assumptions in detail. Already, though, we can see how our simple theory diverges from familiar contemporary theories. Since Partee and Rooth (1983), it has been standard to abandon **Inflexibility** by allowing type-shifting. The idea is that an expression may make different semantic contributions, depending on whether a type-shifting principle is applicable.[12]

---

11 This example is just a toy. There are all sorts of complexities that may affect which type we assign to the meaning of *is swimming*; I ignore them here.
12 There's a complication: one may reasonably regard type-shifting principles as shifting between different meanings rather than shifting single meanings between types. This complication won't matter for my purposes. What's important is that our simple theory disallows flexibility. Partee (1986) discusses the status of type-shifting principles, though, as noted by Geurts (2006), there are a number of different views one may have about such principles.

Now we can give a type-theoretic semantics for a tiny fragment of English. Our fragment contains the atomic predicate *wise*, and the name, *Frege*. The semantic value of *wise* is type <e,t> while the semantic value of *Frege* is type <e>.

[[*wise*]]= λx.x is wise.

[[Frege]]= Frege

In accordance with our single composition rule – function application – we can now also interpret *Frege is wise,* which is the result of applying the function designated by *wise* to the entity designated by *Frege*. Since [[wise]] is a function from entities to truth values, [[*Frege is* wise]] is type <t>, i.e. it is true or false. Until I specify otherwise, I will work with an extensional semantics for our fragment.

In giving this interpretation, we've assumed that the copula is semantically vacuous. This is particularly controversial when discussing the semantics of predication, as it is natural to take the copula to be a significant part of the predicate, perhaps even the part that enables predication. We'll return to this.

Vacuity: the copula is semantically vacuous.

Expressibility problems arise when we use these same type-theoretic methods to analyze more complicated fragments of English. Problem **MF** was that Frege's semantic theory prevents us from adequately analyzing statements about semantics, problem **NF** was that Frege's semantic theory prevents us from adequately analyzing natural language constructions that seem to require co-reference between predicates and non-predicates.

The relevant versions of **MF** and **NF** are as follows. **MF'**: our simple type-theory cannot adequately analyze all of the sentences necessary in order to express semantic theories. **NF'**: our simple type-theory cannot adequately analyze natural language predicate nominalization.

Both **MF'** and **NF'** utilize the elusive notion of adequate analysis. For the sake of this discussion, a skeletal view about semantic analysis will suffice. In particular, we'll take a semantic theory to provide an adequate analysis of a sentence only if it assigns semantic values to its sub-sentential constituents, along with general principles deriving the semantic value of the sentence built from those constituents, such that the assigned semantic value (i) truth-conditionally matches the analysandum, and (ii) captures the meaning of the analysandum. Fully understanding requirement (ii) is difficult, but some violations are obvious. For instance, if our theory takes *Snow is white* to express that snow is white and grass is green, this violates (ii). The constraint is familiar from Davidsonian meaning-theory construction.[13]

To make a case for **MF'** and **NF'** I'll consider two example sentences:

(12) *Wise* designates wisdom.
(13) Wisdom is a property that Frege has.

The problem comes in attempting to analyze (12) and (13) with the resources available. Our extremely simple theory is inadequate for an uninteresting reason: both (12) and (13) contain vocabulary that goes beyond our tiny lexicon (which, thus far, only contains two entries). The first thing for a type-theorist to do, then, is to give interpretations of these other elements. In other words, we need to extend our lexicon so that we can interpret *"wise"*, *designates, wisdom*, and *a property that Frege has*.[14] Giving interpretations for each of these words requires two things: (a) identifying the type of their meaning and (b) identifying their semantic value, i.e. exactly what their semantic value is.

Before tackling (a) and (b), note that to draw out **MF'** and **NF'** I am assuming that the same type-theoretic methods we use to analyze *Frege is wise* can be extended to (12) and (13). This is not uncontroversial: it is familiar for theorists to claim that their semantic analyses only apply to a certain subset of linguistic expressions.

No relevant limits: the same type-theoretic methods we use to analyze *Frege is wise* can be extended to analyze (12) and (13).

Let's now tackle (a). Here's a constraint on type-assignments: they should cohere with our already assigned types in order to make correct predictions about interpretability. Given that we analyzed *wise* as having an <e,t>-type meaning, and *"wise" is wise* is interpretable (though obviously false), we need to take *"wise"* to be assigned a meaning that can combine with an <e,t> meaning via function-application. There are only two options: <e> and <<e,t$>,$t>.[15] Noting that *"wise"* appears to have the same sort of distribution as a proper name makes it plausible that its meaning is <e>-type. Using analogous reasoning for the other expressions, we end up with the following categorizations (using ∈ to signify category membership). On these category assignments, (12) and (13) are perfectly interpretable, as desired. I'll ignore the internal complexity of *a property that Frege has*.

---

13 See Lepore and Ludwig (2007) for discussion.
14 Though I'm primarily using italics to mention expression, I use quotes when I need to mention a mention, as in *"wise"* which refers to the word *wise*.
15 As a matter of fact, none of the arguments I give will depend on assigning *"wise"* and <e> meaning rather than an <<e,t>,t> meaning.

[["wise"]] ∈ <e>
[[designates]] ∈ <<e>,<e,t>>
[[wisdom]] ∈<e>
[[*a property that Frege has*]] ∈<e,t>

These categorizations are not uncontroversial. One may, for instance, argue that [[*wisdom*]] should be type <e,t>. I will consider such a suggestion, in effect, in both 3.1 and 3.3. For now, it will suffice to see how versions of **MF** and **NF** arise given the above type-assignments, which are an obvious initial proposal.

To identify semantic values, we can be guided by disquotation. The following are near truisms: *Wisdom* designates wisdom and *designates* designates designation. Combining these observations with the type-assignment we get the following semantic values:

[["*wise*"]] = wise
[[*designates*]] = λx.λy.y designates x
[[*wisdom*]] = wisdom
[[*a property that Frege has*]] = λx.x is a property that Frege has.
[[*Frege*]] = Frege

This reasoning utilizes a link between interpretation (as expressed by the double-bracket notation) and designation, which some may find suspect.

Bridge [[*x*]] = y iff *x* designates y.

To argue that this analysis of (12) fails I'll show that, given the principles articulated, the analysis predicts that (12) is false. (12), however, is true. To give the argument, I'll stipulate that *object* is a predicate true of anything that is type <e> and *concept* is a predicate true of anything that is type <e,t>. Given **Exclusivity**, no object is a concept. We can then give an argument that, according to the analysis, (12) is false:

1. *Wise* designates a concept. (From the definition of *concept* and the designation-assignment for *wise*)
2. *Wisdom* designates an object. (From the definition of *object* and the designation-assignment for *wisdom*)
3. No object is a concept. (**Exclusivity**)
4. It is not the case that *wise* designates what *wisdom* designates. (from 1–3)
5. Therefore, It is not the case that *wise* designates wisdom. (from 4 and the intersubstitution of *what "wisdom" designates* and *wisdom*)

On the proposed analysis, (12) is false. However, (12) is true. Therefore, the proposed analysis fails (i): it lacks truth-conditional adequacy.

Here is an argument that the analysis fails (ii): however (12) is analyzed, it should contain reference to the denotation of *wise*. However, it doesn't: after all it does not contain a lexical item that even has a meaning of the same type, let alone the same denotation. So, the analysis of (12) fails to capture its meaning.

This argument is quick: at least prima facie, one can resist it by denying that capturing the meaning of (12) requires reference to the denotation of *wise*. Familiarly, Montague (1974) divorced the semantic analysis of names from their referents by taking names to be generalized quantifiers. Nonetheless, Montague's analysis doesn't obviously violate (ii). I'll return to this argument in section 3.1.

These arguments have nothing to do with the particular examples chosen. Our simple type theory will generate trouble for any sentence purporting to ascribe a non-<e>, or <<e,t>,t>-type meaning to an expression. This is due to the fact that *designates* seems to require an expression in its object position that is either type <e>, or <<e,t>,t>. In all other cases, there will be a mismatch between the type of the word mentioned in the subject position of a designation-clause and the type of the word in the object-position. The generality of the problem is the basis for concluding that our simple type-theory falls prey to **MF'**: the inability to adequately analyze statements ascribing denotation.

To argue that the proposed analysis of (13) fails, I'll make two claims about (13), both concern their relationship to (14):

(14) Frege is wise.

The first is that (13) and (14) have the same truth-conditions. The second is that this holds in virtue of the meanings of the words that compose (13) and (14), along with an uncontroversial claim about property-instantiation (I'll return to this). The contrast is with pairs of sentences like *two plus two equals four* and *three times three equals nine* which have the same truth-conditions but not in virtue of meaning or form. An adequate semantic theory should underwrite the truth-conditional equivalence of (14) and (13), but the proposed semantic theory does not. Nothing in our simple semantic theory connects the semantic values of *wise* and *wisdom* they are members of different types, which cannot overlap. So, we now have an argument that our simple analysis of (13) fails (i): an analysis of (13) is truth-conditionally adequate just in case it ensures that (13) is equivalent to (14); our analysis does not; therefore, our analysis is not adequate.

The standard solution to this worry is to enrich the analysis with some sort of mechanism that ensures the truth-conditional equivalence of (14) and (13): e.g. by taking wisdom to be somehow correlated with λx.x is wise. I will critically discuss such approaches in section 3.3. Note, though, that this is not the only solution. If we abandon **Exclusivity**, we can claim that $[[wisdom]] = [[wise]]$ while maintaining that $[[wisdom]] \in \langle e \rangle$ and $[[wise]] \in \langle e,t \rangle$; this is because meanings can be members of more than one type. In and of itself this does not guarantee that (14) and (13) are truth-conditionally equivalent, but it does link them in such a way that has the potential to undergird such an explanation. I'll return to this in section 3.4

Just as we gave a crude argument that our simple theory fails (ii) in its analysis of (12), we can give a similarly crude argument that our simply theory fails (ii) in its analysis of (13). (13) and (14), the thought goes, both ascribe wisdom to Frege. This seems to require that they both contain constituents that designate Wisdom. Assuming the candidates are *wise* and *wisdom*, and that *wisdom* designates wisdom, only (13) is about the right property.

This crude argument, like the one that came before it, relies on linking designation with subject-matter (aboutness). These links are controversial, and with the exception of Lewis (1988) and Yablo (2014), there are few worked-out theories of subject matter and none which would straightforwardly link designation and subject matter. Even so, the arguments capture the idea that type-distinctions force us to divorce semantic values that, intuitively, coincide.

3. Attempted Solutions

Our extremely simple type-theory yields problems **MF'** and **NF'**. This, in and of itself, may be unsurprising: after all, the simple theory is close to Frege's own – at least at the level of reference. More interesting and surprising results come from the examination of complications to the simple theory that may seem to solve the problems. In particular, the abandonment of **Inflexibility** may seem particularly promising given that, as a matter of fact, almost all semanticists endorse some sort of flexible typing. I'll argue, though, that abandonment of **Inflexibility** – or, for that matter, **Vacuity**, **Bridge**, or **No relevant limits** – do not solve the problems. My preferred solution requires abandoning **Exclusivity**, though it too faces empirical and philosophical challenges.

3.1 Interpretation, Denotation, Meaning, and Reference

In informally discussing semantic theories, it is tempting to take the interpretation assigned to an expression in a theory as the denotation or meaning of that expression. In our discussion, this tendency has already manifested itself. When assigning semantic values, we were guided by **Bridge**, which explicitly links interpretation and denotation.

Bridge: [[x]] = y iff x designates y.

Bridge, however, is suspect, along with other principles linking interpretation with denotation, meaning, and reference. We can generate doubts about these principles both by considering a specific example from Montague, as well as reflecting on the aims of a semantic theory.

Principles like **Bridge** are suspect because there is a plausible view on which interpretation is divorced from denotation, meaning, and reference. On the view, interpretation is a theory-internal notion. Take a given set of goals for a semantic theory and conjoin them, the interpretations of some expressions relative to those goals are the objects that play the relevant functional role in satisfying those goals. Imagine that the only goals of a semantic theory are truth-conditional adequacy and compositionality. In that case, interpretations would be whatever jointly satisfied those constraints.

Contrastively, denotation, meaning and reference are naturally taken to be theory-external notions. The idea is that we utilize and grasp these notions independently of our specific goals in constructing semantic theories, so we cannot take them to be whatever happens to satisfy a theoretically-defined functional role. For instance, *Frege* refers to Frege, and this fact seems independent of whatever goals we have in constructing a semantic theory.

One could accept this picture and argue for principles like **Bridge** by linking theory-external notions like reference with theory-internal notions like interpretation. This would require adopting a particular set of theoretical goals and arguing that, as a matter of fact, those goals ensure links between interpretation and denotation. Fortunately for us, we needn't go down this road: as I'll soon argue, versions of **MF** and **NF** arise even without linking interpretation to reference, denotation and meaning.

Another worry for **Bridge** arises from the fact that a familiar theory violates it. Montague (1974) claims that all noun-phrases are type-identical: they all are interpreted as generalized quantifiers (<<e,t>,t>). [[*Frege*]] for instance, is a function from properties that Frege instantiates to truth, and all others to falsity. However, one may still reasonably insist that *Frege* denotes Frege. This is just a case where interpretation – semantic value – doesn't perfectly track denotation. In fact, one can even understand Montage as taking semantic value to be subject to independent constraints. In this case, the constraint is that expressions in a syntactic category are analyzed as semantically type-identical.

The upshot is that **Bridge** has been rejected, and its rejection can be reasonably supported by distinguishing semantic interpretation from more familiar notions. It may, then, be tempting to think that problems **MF** and **NF** arise due to endorsement of **Bridge** and other similar principles.

This temptation should be resisted. Versions of **MF** and **NF** arise even if we reject **Bridge**, or any other similar principles. To see that **MF** arises without **Bridge**, begin by substituting *is interpreted as* for *denotes* in our argument for the claim that the analysis of (12) was inadequate.

1) *Wise* is interpreted as a concept. (From the definition of *concept* and the designation-assignment for *wise*.)
2) *Wisdom* is interpreted as an object. (From the definition of *object* and the designation-assignment for *wisdom*.)
3) No object is a concept. (**Exclusivity**)
4) It is not the case that *wise* is interpreted as what *wisdom* is interpreted as. (1–3)
5) Therefore, It is not the case that *wise* is interpreted as wisdom.

Now, consider an interpretation-specification for *wise* that may seem to support the premises of the revised argument:

(15) *Wise* is interpreted as wisdom.

The idea would be that (15) is interpretable, and, in order to make it so, we'd have to assign *wisdom* type <e>. If that's correct, then the argument could proceed as before. However, there's a problem. We're considering a view on which the notion of interpretation is theory-internal and plays whatever role we need it to. So, we're not licensed in taking *wisdom* to be of type <e> until we know more about this role. For all that has been said, the initial proposed modification of the argument for **MF** may fail given the theory-internal conception of interpretation.

However, a novel argument can be generated utilizing these observations. Consider (15) and (16), both of which utilize the theory-internal notion of interpretation.

(16) *Frege* is interpreted as Frege.

Now consider attempting to analyze *is interpreted as* using our familiar methodology. *Wise* is a first order monadic predicate so, by hypothesis, it will be analyzed as having semantic type <e,t>. *Frege* is a name, so it will be analyzed as having type <e>. The problem arises in analyzing *is interpreted as*. On the one hand, it seems to express $\lambda x.\lambda y.y$ is interpreted as x, where x and y range over entities of type e. This is the interpretation required for (16). On the other hand, it seems to express $\lambda f.\lambda y.y$ is interpreted as f, where y ranges over entities of type e and f ranges over functions of type <e,t>. However, by **Exclusivity** and **Inflexibility** these are different functions and *is interpreted by* cannot express them both. So, **MF** rears its head at another level: our theory fails to allow us to adequately analyze sentences that express interpretation. Given that these are often taken to be the basic clauses in a compositional semantic theory, this is highly problematic.

I have explicated this problem using the English *is interpreted as*. One may worry that this phrase has little to do with the usual construction of semantic theories, which utilize double-bracket notation. However, the problem has nothing to do with the means we use to express interpretation. If we adopt the usual double-bracket notation, the lesson is that our type-theory is inadequate for understanding the meaning of the double brackets. Importantly, this holds both for our simple type theory and our revised version that rejects **Bridge**.

Faced with this argument, a theorist may insist that their theory was never intended to apply to the double brackets. Rather, the theory targeted only natural language, or a small fragment thereof. Note that, independently of scope of the theory, the forgoing reveals that, absent enrichment, our simple type theory cannot extend its scope to cover interpretation specifications. This is quite different from the claim that the theory does not analyze them. I'll return to discussing such potential empirical limitations when I consider the rejection of **No relevant limits**.

Another familiar maneuver is to insist that either *interprets* or the double-bracket notation is syncategorematic. In order to make sense of this view, its proponent owes an account of the meanings of sentences in which the notation occurs. It is hard to see how this could be achieved, though I don't have a principled argument for impossibility. Absent a concrete proposal, though, we have no reason to take a syncategorematic approach to be viable.

Stepping back, the lesson is straightforward. Interpretation, as we usually think of it, is a relation between any expression and its semantic value. Semantic values, however, occupy different places in the type-theoretic hierarchy. This means that interpretation must likewise be able to relate expressions with entities in different locations on the hierarchy. However, if meanings are exclusive and inflexible, there can be no such meaning. So, there is no adequate manner to analyze any expression that purports to express the interpretation relation.

At this stage a theorist could simply insist that there is not a single interpretation function and the double-bracket notation is ambiguous between relations for every type (or at least every type attested in our target language). This would preclude comparing semantic values, e.g. *"Frege" is interpreted as an object, while "wise" isn't*. This would also preclude the straightforward statement of composition rules, like the one given above – our statement of function application crucially utilized double-bracket notation. On the theory envisioned, that notation is many-ways ambiguous, so it would need to be disambiguated. We would then be left with myriad composition rules, just as we had myriad interpretation functions. While this isn't incoherent, it would be resisted by most theorists who take there to be very few composition rules.

Again, I won't offer anything like a decisive consideration against taking the double bracket notation to be infinitely ambiguous. I'll simply offer some skepticism and the observation that it would be quite surprising if this infinite ambiguity – and the attendant theoretical complications – were the inevitable result of attempting to generalize type-theoretic analysis to semantic discourse.

What about problem **NF'** and **Bridge**? Here matters are more straightforward. Merely abandoning **Bridge** does nothing to ensure that the truth-value of (13) matches the truth-value of (14). The initial problem was that our semantic theory did nothing to guarantee that match, and weakening the theory certainly doesn't help!

As straightforward as this argument is, it illuminates the shortcomings of rejecting **Bridge**. The motivation for rejecting **Bridge** was the thought that concept horse problems arise by misconstruing semantic interpretation as intimately related to more familiar notions. The mistake is brought out by realizing that the mutual entailments between (14) and (13) have nothing to do with how we construe interpretation. Insofar as these entailments are within the purview of a semantic theory, then however we understand interpretation, the entailments will have to be guaranteed by our interpretations of (14) and (13).

3.2 The Copula

Thus far, we've been assuming that the copula is semantically vacuous. This can be contested for a number of reasons. Most obviously, one could take the copula to semantically mark the present. For our purposes, however, the more important worry is that *wise* is not itself a predicate: it only becomes a predicate by combining with the copula.[16] The function of the copula, in turn, can be understood as transforming a referential general term (that refers to a property or kind) into a predicate. On this view, **Vacuity** is rejected.

Vacuity: the copula is semantically vacuous.

---

16 Wiggins (1984) defends this view.

Rejecting **Vacuity** is unhelpful for two reasons. The first is that it is empirically suspect, and the second is that it does not solve problems **MF'** and **NF'**.

If *wise* can serve as a predicate without the copula then the copula is not always responsible for predication. In fact, there are occurrences of *wise* in which it can serve as a predicate without the copula.

(17) I consider Frege wise.

In (17), the complement of *consider* is the small clause *Frege wise*. Here, despite the absence of the copula, *wise* is functioning as a predicate: it ascribes wisdom to Frege.[17] In order to account for examples like (17), proponents of the view that the copula is the locus of predication must claim that such sentences contain tacit constituents performing the same semantic function that they take the copula to perform (14). In fact, there may be some support for this view, at least in sentences like (17). Bowers (1993) and Baker (2003) argue that sentences contain a tacit Pred-operator that has the semantic function of creating predicates. Whether such a hypothesis is plausible in all cases is an open issue. So, at the very least, those who take *wise* to be a non-predicate that is transformed into a predicate by some other linguistic item are making a controversial empirical hypothesis.

More importantly, even if the predicate in (14) is *is wise* rather than *wise*, problems **MF'** and **NF'** remain. Taking **NF'** first, note, again, that taking the copula to be the locus of predication again does nothing to guarantee truth-conditional match between (13) and (14). Regarding **MF'**, note that we can simply substitute *is wise* for *wise* in our initial argument for **MF'**.

3.3 Type-Shifting

In deriving **MF'** and **NF'** I assumed **Inflexibility**: that semantic typing is fixed. This is now routinely rejected. Just about every theorist thinks that at least some words may be flexibly typed, in the sense their type is shifted within a particular context, usually in order to preserve interpretability. Given that flexible typing is part of the standard toolkit of semanticists, it is natural to utilize it in accounting for sentences like (12) and (13).[18]

---

17 Rothstein (2004) take small clauses to be the ideal construction for studying predication, her idea being that they do not contain other potentially distracting syntax/semantics.

18 Type-shifting has been prominently used to account for kind-designating NPs and plurals, among many other types of expressions. See Winter (2007) for a recent discussion and references.

The relevant question is whether type-shifting mechanisms can be utilized to provide adequate analyses of (12) and (13). I'll now argue that they cannot for two reasons. The first reason is empirical: no independently-motivated type-shifting principle is suitable for the job. The second reason is conceptual: even if a type-shifting principle could provide adequate truth-conditions it would give rise to problems just as severe as those it was invoked to solve.

Before I argue that no independently-motivated type-shifting principle can provide adequate truth-conditions for sentences like (12) and (13), it is worth explaining why this conclusion is important in the first place. A skeptic, for instance, my claim that independent-motivation is irrelevant, as solving the problems at hand is motivation enough![19]

The reason it is illuminating to seek independently-motivated type-shifting principles is that, without constraint, we could introduce a type-shifting principle to generate whatever truth-conditions we wish by reverse-engineering the principle from the desired truth-conditions. Such an engineered type-shifting principle leads to two worries. The first worry is that allowing this method would result in a gerrymandered and ad hoc semantic theory. There should be some constraint on invoking type-shifting, and the sort of independent motivation to which I allude is a plausible constraint. The second worry is that each type-shifting principle brings with it a risk of overgeneration. For instance, if we can freely shift <e,t> meanings to <e> meanings – without any limitation – we'd predict that any complex expression containing two words with <e,t> meanings is interpretable and can express a truth-value. This prediction, however, is implausible: *wise rich* can no more express a falsehood than any arbitrary combination of like-typed expressions, e.g. *all some*.

There is certainly more to be said about whether the proponent of a type-shifting approach to **MF'** and **NF'** can reasonably introduce a novel type-shifting mechanism. For now, suffice it to say that the fact that no independently motivated principle does the job is a significant strike against a type-shifting approach.

The initial analyses of (12) and (13) failed (ii) because, on the analyses, their sub-sentential constituents didn't designate to the proper entities. Focusing on (13), the problem was that *wisdom* referred to something of type <e>, while, in order for it to be an adequate analysis, it must designate something of type <e,t>. Arguing that (12) fails (ii) is more straightforward than arguing that (13) fails (ii) since the semantic function of predicate nominalizations is complicated and

---

19 Thanks to Dag Westerståhl for pushing this.

controversial.[20] Nonetheless, in what follows I'll focus on (13) for the simple reason that it makes the problems more vivid. If one is convinced that an alternative account of nominalization is superior, note that the same problems arise with (12), they are just a bit more complicated to discuss.

Just how could type-shifting mechanisms help? Recall that our initial type assignment for *wisdom* was dictated by the constraint that we make correct predictions about interpretability. These predictions, in turn, were constrained by the inflexibility of our type assignments. Whatever type is assigned in the lexicon is the only type available for interpretation. Type-shifting principles relax this constraint. A term can have a particular type in a lexicon and appear in an interpretable structure as long as a type-shifting mechanism is available to ensure interpretability. That means that we can assign *wisdom* type <e,t>, and give it the following semantic value.

[[wisdom]] = $\lambda x.$ x is wise

This semantic entry for *wisdom* initially seems superior. After all, *wisdom* is assigned the same semantic value as *wise*, which is what our initial intuitions about nominalization dictated.

Now, consider our analysis of (13). It consists of two sub-sentential constituents, both of which are of type <e,t>. Therefore, we cannot apply function-application. Rather, a type-shifting principle must shift the type of one of the constituents. The first question is which? Insofar as there are such mechanisms at all, they must be available in the nominal domain. The reason is that we can freely conjoin predicate nominalizations with definite descriptions, and combine the conjunctive NP with a single predicate, as in (18).

(18) Wisdom and the number four are both abstract entities.

In fact, there has been a substantial literature regarding type-shifting within the nominal domain, so we are in a good position find candidate mechanisms that suffice. Before critically surveying such mechanisms, let me express some general worries about type-shifting mechanisms.

First, there is the overgeneration worry I already mentioned: if our hypothesized type-shifting mechanism operates whenever we attempt to compose two

---

20 Moltmann (2013) is one recent example of a defence of a view on which nominalizations do not co-refer with their corresponding predicates. Note, however, that even if Moltmann is correct about particular predicate nominalizations, it remains plausible that there are certain argument-type expressions that co-refer with predicates. Property names that are stipulated to so-refer are a particularly vivid example.

expressions with <e,t> meanings, then we'd predict that any complex expression formed of such entities is interpretable. The natural thing to do is to try and limit the application of the mechanism. However, a limitation can't be type-driven given that the mechanism operates in some cases of composition with the same types. Thus, it looks like our quest for type-driven interpretation straightforwardly fails. Type-assignment, and thus, the semantic values of occurrences of terms, are a product of additional considerations.

Second, there is a philosophical worry. Type-shifting mechanisms were intended to help by allowing us to claim that *wise* and *wisdom* co-designate. However, given that a type-shifting mechanism shifts the type of *wisdom* in (13), it is hard to see just how, even on such a view, (13) satisfies (ii). After all, the meaning ultimately utilized in calculating the truth-conditions of (13) will not be that of *wisdom*, but, rather, a shifted variant. In order to argue that (ii) is satisfied, these variants must be closely connected to their correlative unshifted meanings.

On to the particular type-shifting mechanisms. There are two options: we can shift *wisdom* to the higher-type <<e,t>, t>, or the lower type <e>. In fact, both mechanisms have been hypothesized to exist in the nominal domain. Let me first focus on the mechanisms for shifting *wisdom* to a higher-type. These mechanisms fail (i): they deliver incorrect truth conditions for (13).

Here are two standard mechanisms for shifting from <e,t> to <<e,t>,t>:

$A(P_{<e,t>}) = \lambda Q_{<e,t>} \exists x(Px \wedge Qx)$

$THE(P_{<e,t>}) = \lambda Q_{<e,t>} \exists! x(Px \wedge Qx)$

If we shift [[*wisdom*]] using A, the truth conditions of (13) is taken to be true just in case a wise thing is a property of Frege. If we shift [[*wisdom*]] using THE, (13) is true just in case the wise thing is a property of Frege. Both of these truth-conditions are incorrect.

The more promising type-shifting strategy shift from <e,t> to <e>. Intuitively, such a type-shift will produce an entity correlate for each property. This mechanism is motivated by the existence of kind-designating occurrences of plurals and mass nouns.

(19)   Dinosaurs are extinct.
(20)   Water is plentiful.

If we take *dinosaurs* and water to be assigned <e,t>-type meanings in the lexicon, then we'll plausibly need to shift them in order to interpret (19) and (20). (19), for instance, expresses that a particular kind of thing is extinct. These constructions motivate the existence of an <e,t> to <e> type-shifting mechanism that maps properties to kinds.[21]

The first reason that this type-shifting mechanism does not help is that it does not yield truth-conditionally adequate interpretations. We can see this once we are a bit more careful about the ontological distinctions between kinds and properties.[22] Kinds, it seems, have features more closely tied to their instances. Kinds can go extinct, be invented, evolve, etc. Properties, however, don't seem to have such properties. They are ordinarily conceived-of as necessarily existing entities that cannot change in these ways. Nominalization and interpretation-specification seem to require reference to properties – the semantic values of predicates – rather than kinds. So, it seems as if the <e,t> to <e> type-shifting mechanism invoked to make sense of kind-reference will not generate the correct truth conditions for nominalization or metalinguistic discourse.

An adequate understanding of the distinction between kinds and properties is far beyond the scope of this discussion. Rather than further pursuing the issues, I'll now turn to the conceptual problem for a type-shifting approach to **MF'** and **NF'**.

Even if we could identify a type-shifting principle that gives rise to adequate truth-conditions for (13) and (12), problems remain. Invoking such type-shifting principles gives rise to revenge problems, which are just as severe as the problems they set out to solve. Let us introduce the name *Sam* with the stipulation that its interpretation is the type-shifted interpretation of *wisdom*. Using *Sam* to discuss the proxy-object itself, we will want to claim that the following is false:

(21) Sam is a concept.

The problem is that according to the semantic version of the strategy under consideration, (21) is true. *Sam*, by stipulation, co-designates with *wisdom*. As such, intersubstitution of the two will not affect the thought expressed. Since, by hypothesis, *Wisdom is a concept* is true – after all *wise* designates wisdom, and *wise* designates a concept – then *Sam is a concept* will also be true, given that the two express the very same thought. This reasoning generalizes to every context in which we wish sentences intuitively about the objects that result from type-shifting themselves to truth-conditionally diverge from sentences in which those objects are playing the role of mere representatives.

---

21 See Chierchia (1998) and (2011) on such mechanisms.
22 See Liebesman (2011) on the differences between kinds and properties.

Solving this problem would require distinguishing Sam in its role as the referent of *Sam* and Sam in its role as the type-shifted meaning of *wisdom*. This, however, requires much more than type-shifting. It requires understanding how a single meaning can play two different roles in two different occurrences. In fact, my favored view, which rejects **Exclusivity** does make sense of this. However, once we adopt that view, there is no need to invoke type-shifting principles.

The upshot of all of this is that no independently motivated type-shifting principles can allow us to adequately analyze (12) and (13) and that we have a principled reason to think that no type-shifting principle could be adequate. The reason is that if a type-shifting principle provides adequate truth-conditions it seems to give rise to an expressibility problem just as bad as **MF'** or **NF'**: we cannot provide adequate truth-conditions for sentences about the type-shifted meanings themselves.

3.4 Empirical Limitation

According to **No relevant limits**, we can straightforwardly apply our methods of type-theoretic analysis to (12) and (13). This can be reasonably doubted. Perhaps, as has been commonly held since Tarski, we should hold that we cannot utilize our methods to apply to the metalanguage itself. Rejection of **No relevant limits** can be evaluated as a strategy for undermining **MF'**, **NF'**, or both. I'll address them in turn.

Begin with **MF'**, where rejecting **No relevant limits** seems well-motivated, given that **MF'** explicitly deals with metasemantic discourse. Rejection of **No relevant limits**, however, doesn't solve a variant of **MF'**, which requires no analysis of metalinguistic vocabulary. The idea is that we can directly argue that, for the type-theorist, it is false that *thinks* designates thinking. Here's the argument:

(1) *Thinks* designates something of type <e,t>.
(2) *Thinking* designates something of type <e>.
(3) Nothing of type <e> is also of type <e,t>.
(4) Therefore, it is not the case that *thinks* designates thinking.

Note that this argument does not require applying our type-theoretic methods to any metalinguistic vocabulary. The support for the premises of the argument does not come from some particular analysis of those premises or the terms used in their articulation, rather, the motivation comes from the tenets of the type-theory itself. Another way of putting this is that the type-theorist will endorse the premises even if he/she rejects the possibility of analyzing *designates* with type-theoretic methods. So, the rejection of **No relevant limits** is of no help. Furthermore, the problem can be easily generalized to any specification of predicate designation (interpretation/reference/etc.) that purportedly utilizes a non-predicate to specify that designation. This shows that merely restricting our type-theoretic analyses such that they fail to cover metalinguistic vocabulary does not solve the underlying problem.

When it comes to **NF'**, the rejection of **No relevant limits** doesn't help at all. The reason is that **NF'** does not concern metalinguistic notions. Rather, it concerns any non-predicate that, intuitively, co-designates with predicates. These are perfectly familiar natural language expressions like gerunds and infinitives. To argue that type-theoretic methods cannot be used to understand gerunds and infinitives is to take there to be limits to our type-theory that aren't motivated by the object language/metalanguage distinction.

Given these shortcomings, I'll set aside the rejection of **No relevant limits** as a method for solving MF' and NF'. A second reason to do this is that, as far as I know, no actual theorists have approached the problems in this way.

3.5 Type Overlap

According to **Exclusivity**, no meaning can be a member of more than one type. This is precisely what forced us to take *wisdom* and *wise* to have different meanings. The reasoning was that *wisdom* has an <e>-type meaning while *wise* has an <e,t>-type meaning, and since no meaning can be a member of both types, the meanings must be distinct. If we abandon **Exclusivity**, we can reject this reasoning and claim that the terms have the same meaning, though the former has it in an <e>-ish way while the latter has it in an <e,t>-ish way.

Abandoning **Exclusivity** raises a number of questions. First, how can we make sense of non-exclusive types? Second, what is it to have a meaning in a way? Third, what becomes of type-driven interpretation? Fourth, what does an **Exclusivity**-denying compositional semantics look like? Fifth, how do we ensure that the resulting theory does not succumb to the property version of Russell's paradox. I won't address the last question – which is a topic for another paper (book?). In the remainder of this subsection, I'll address the first two questions and in the last section of the paper I'll turn to the third and fourth.

We can begin to make sense of non-exclusive types by considering categorization more generally. I can divide objects into categories in a number of ways. For instance, my colleagues could be categorized by height: those above 6' tall are in one category and the rest are in the other. This would yield categories with no overlap. I could also categorize my colleagues by their taste in music with categories for jazz, classical, rock, and EDM. This categorization would allow overlap: lots of jazz lovers also like classical music. In this latter case we can make sense of overlapping categories by making sense of the principle on which the categorization is based.

Making sense of overlap in the type-system is similar. Overlapping categories can be made intelligible by providing a principle on which an overlap-allowing categorization could be based. The principle will be motivated by consideration of the role of types. Following Frege, I will take typing to be determined by function/argument structure. Recall that Frege took this to be the key to understanding the unity of thoughts. Of course, Fregean thoughts are composed of senses and we've been considering an extensional semantic theory thus far, so we need to transpose his ideas to the present context. We can begin to understand semantic typing by posing questions about the semantic values of sentences. Unlike Frege, the contemporary type-theorist often does not take structured entities to be the semantic values of sentences. So, we cannot pose unity questions in terms of the structure of propositions/thoughts. However, we can ask a similar question that does not presuppose that sentence-level semantic values are structured.

> **Q**: What is the nature of the relation that obtains between some entities in virtue of which they generate a proposition?

Q lacks presuppositions about the structure of propositions, so it is compatible with the standard view among contemporary type-theorists that propositions are unstructured sets of worlds.[23] The idea is that in answering Q, we can make use of distinctions among semantic values.

In posing **Q** we are taking sentence-level semantic values to be propositions. So, we're moving to the intensional. Of course, taking the extensions of sentences to be truth-values, we can pose an analog of **Q** that asks what relation obtains between some entities in virtue of which they generate a truth-value.

One may be tempted to say that the answer to **Q** is entirely straightforward: function-application. However, in the present context a full answer to **Q** requires not merely naming the relevant relation but characterizing it – giving its nature – this, in turn, will require characterizing its adicity and relata.

We can distinguish between **Q** and what I'll call *instances of* **Q**, which ask what enables specific groups of entities to generate a proposition. Also, we shouldn't take *generate* particularly seriously, lest we be saddled with a rejection of the view

---

23 It would take a bit more work to find a similar question that doesn't presuppose the existence of sentence-level semantic values at all. However, it can be done. We can ask, for instance, what relationship there must be between some meanings such that the sentence that expresses them is truth-apt.

that propositions necessarily exist. Rather, we should focus on the distinction between groups of entities that correspond to propositions and those that don't. I'll use *generate* to cover the former groups.

An example makes this vivid. Suppose we want to answer **Q** for a specific set of intensional semantic values of constituents of *Frege is wise*. In that case, we'll invoke the nature of the function [[*wise*]] along with the nature of the object, [[*Frege*]], and finally, a relation between them: that [[*Frege*]] is the argument of [[*wise*]].

From this example, we can extract categories that are important for answering **Q**. The category of being able to play the role of an argument is one category. The category of being able to play the role of a propositional function is another. My view is that some entities are in both of these categories. The rejection of **Exclusivity** follows.

In sum, categorization is – at least to a first approximation – determined by the explanatory roles invoked in answering **Q**. In our type-theory, these are just the familiar function/argument roles handed down by Frege. My thesis is that single entities can play different roles in different circumstances. This is intuitive: it seems as if wisdom plays a predicative role in the meaning of *Frege is wise* but a non-predicative role in the meaning of *Wisdom is a virtue*.

All of this answers our first question: we can make sense of type-overlap in terms of the explanatory role of types. Categorizing entities in terms of answers to instances of **Q** allows that an entity can play one role in one answer, and another role in a different answer. We can now move to the second question: what is it to have a meaning in a way?

Crucial to the view being developed is the thought that the same entity can play different roles. Which role is played by an entity is, in turn, determined by the way in which that entity is picked out by its corresponding linguistic item. Again, in *Frege is wise*, *wise* picks out wisdom in a predicative way, whereas in *Wisdom is a virtue*, *wisdom* picks out wisdom in a non-predicative way. These ways are simply different designation relations. *Wisdom* refers to wisdom, while *wise* ascribes that property to Frege. Making sense of the view requires making sense of these different relations – reference and ascription.

In Liebesman (2015), I articulate, develop, and defend two contrasts between reference and ascription. The first distinction between reference and ascription is that the latter is fundamentally triadic whereas the former is fundamentally dyadic. The idea is that occurrences of predicates ascribe properties to their arguments whereas occurrences of non-predicates simply refer. Notice that this distinction takes occurrences to be more semantically fundamental than words themselves, at least in terms of answering **Q**. The second distinction between reference and ascription is that the former is wholly unconstrained – any entity can be a referent – while the latter is constrained. Only predicables, most familiarly properties and relations, can be ascribed to things.

We can distinguish reference and ascription by their adicity and relata. However, we can also distinguish them by considering the different roles they play in answering **Q** and its instances. This, in turn, is displayed by incorporating them into a compositional semantic theory, which is the topic of the next section.

4. Sodium-Free Semantics

Rejecting **Exclusivity**, along with distinguishing multiple semantic relations, sits uneasily with familiar type-theories. Not only do they standardly make use of just one interpretation relation, but composition is type-driven in the sense that function-argument structure is determined by the types of semantic values being composed. If a single semantic value can belong to more than one type, we cannot take composition to be type-driven. This problem is exhibited by (22) and (23).

(22) Being nice is intelligent.
(23) Being intelligent is nice.

On the theory under consideration [[*being nice*]] = [[*nice*]] and [[*being intelligent*]] = [[*intelligent*]]. Given this, the meanings of (22) and (23) are derived by composing the same meanings. If composition is wholly type-driven, it seems that we have no means to distinguish the semantic values of the sentences. What we need is some way to ensure that [[*nice*]] plays an argument role in (22) and a predicate role in (23). My next step is to sketch a semantics that allows us to make sense of this.

We can begin by incorporating into our semantics the claim that there are multiple interpretation relations. In particular, we'll take there to be three interpretation functions. The first two are are partial: ref and asc. The former, intuitively, is for argument-type expressions while the latter is for predicative type expressions. So, ref(*being* nice) = asc(*nice*). We'll also continue to use double-bracket notation for the more general interpretation relation, which, so understood, is a deteminable that has ref and asc as its determinates. Composition, in turn, is sensitive to whether an expression is in the domain of ref or asc. To capture this, we add the following composition principle, where dom is is a function from a relation to its domain:

If w ∈ dom(ref) and y ∈ dom(asc) then [[wy]] = [[y]]([[w]])

This principle allows us to interpret simple atomic sentences in which we have a single predicate and a single non-predicate. So, for instance, (22) is easily interpretable because *being nice* is in dom(ref) and *intelligent* is in dom(asc). The result is that asc(*intelligent*) takes ref(*being nice*) as its argument. The meaning of (22) will then differ from the meaning of (23) because the latter has the reverse function/argument structure: ref(*being intelligent*) is the argument of asc(*nice*).

This view works well for familiar atomic sentences, but hits a snag as soon as we consider sentences that contain higher-order predicates.

(24)   All dogs bark.

In order to interpret (24), we wish to compose the meanings of *all* and *dogs*. There's a problem: they both seem to be predicative expressions and, therefore, are in the domain of asc. Our single composition rule does not allow us to compose their meanings. To make sense of such cases, we introduce a second composition rule that allows us to compose predicates:

If $w \in$ dom(asc) and $y \in$ dom(asc) then $[[wy]] = [[y]][[w]]$

Notice that the composition rule does not specify which of $[[y]]$ and $[[w]]$ is the function, and which is the argument. This is because when we are composing predicates, composition can remain type-driven. In the case of *all* and *dogs*, $[[all]]$ can take $[[dogs]]$ as its argument, but not vice-versa.

That our second rule remains type-driven is a key philosophical contrast between these rules. It is motivated by the following thought: any entity (of any type) can be named and can serve as an argument. We can mark this by the way in which it is picked out; it is in the domain of ref. Predication, however, is limited to predicables. Furthermore, when an entity functions predicatively, it must do so in accordance with its type. This guarantees that interpretation can remain type driven when we combine two predicates.

The theory just sketched is extremely skeletal. A full evaluation of its merits would require the attempt to extend it to a much larger fragment of natural language. However, it should already be clear that it has the merits both of avoiding MF' and NF', and remaining fairly elegant: it just has two composition principles. My conclusion is that rejection of **Exclusivity** allows us to construct a semantic theory that doesn't require us to request, as Frege did, a grain of salt.[24]

---

24   Thanks to an audience at University of Stockholm, as well as Matti Eklund, Ali Kazmi, Mark Migotti, and two anonymous referees.

References

Baker, M. (2003). *Lexical Categories*. Cambridge: Cambridge University Press.

Barker, C. and Jacobson, P. (eds.) (2007). *Direct Compositionality*. Oxford: Oxford University Press.

Bäuerle, R., Schwarze, C. and von Stechow, A. (eds.) (1983). *Meaning, Use, and Interpretation of Language*. Berlin: Walter de Gruyter.

Beaney, M. (1997). *The Frege Reader*. Oxford: Blackwell.

Bowers, J. (1993). "The Syntax of Predication". *Linguistic Inquiry* 24: 591–656.

Burge, T. (2007). "Predication and Truth". *Journal of Philosophy* 104: 580–608.

Burgess, J. (2005). *Fixing Frege*. Princeton: Princeton University Press.

Chierchia, G. (1998). "Reference to Kinds Across Languages". *Natural Language Semantics* 6: 339–405.

Dummett, M. (1973). *Frege: Philosophy of Language*. London: Duckworth.

Frege, G. (1892). "On Concept and Object". In: M. Beaney (1997), 181–193.

Furth, M. (1968). "Two Types of Entity Designated". In: N. Rescher (ed.), 9–45.

Hale, B. and Wright, C. (2012). "Horse Sense". *Journal of Philosophy* 109: 85–131.

Lepore, E. and Ludwig, K. (2007). *Donald Davidson's Truth-Theoretic Semantics*. Oxford: Oxford University Press

Liebesman, D. (2011). "Simple Generics". Noûs, 405: 409–442.

Liebesman, D. (2015). "Predication as Ascription". *Mind*, 124: 517–69.

MacBride, F. (2011). "Impure Reference: A Way Around the Concept Horse Paradox". *Philosophical Perspectives* 25: 297–312.

Moltmann, F. (2013). *Abstract Objects and the Semantics of Natural Language*. Oxford: Oxford University Press.

Partee, B. (1986). "Noun Phrase Interpretation and Type-Shifting Principles". Reprinted in B. Partee (2004).

Partee, B. (2004). *Compositionality in Formal Semantics. Selected Papers by Barbara H. Partee*. Malden, MA: Blackwell.

Partee, B. and Rooth, M. (1983). "Generalized Conjunction and Type Ambiguity". In: R. Bäuerle et al. (eds.), 361–383.

Potter, Michael and P.Sullivan (eds.) 2013: *Wittgenstein's Tractatus: History and Interpretation*. Oxford: Oxford University Press.

Proops, I. (2013). "What is Frege's Concept Horse Problem?" In: M. Potter and P. Sullivan (eds.), 76–96.

Rescher, N. (1968). *Studies in Logical Theory*. American Philosophical Quarterly Monograph Series 2, Oxford: Blackwell.

Rieppel, M. (forthcoming). "Being Something: Properties and Predicative Quantification". *Mind*.

Rothstein, S. (2001). *Predicates and their Subjects*. Dordrecht: Kluwer.

Strawson, P. F. (1974). *Subject and Predicate in Logic and Grammar*. London: Methuen.

Trueman, R. (2015). "The Concept *Horse* With No Name". *Philosophical Studies* 172: 1889–1906.

Wiggins, D. (1984). "The Sense and Reference of Predicates: A Running Repair to Frege's Doctrine and a Plea for the Copula". *Philosophical Quarterly* 34: 311–328.

Winter, Y. (2007). "Type Shifting With Semantic Features: A Unified Perspective". In: C. Barker and P. Jacobson (eds.), 164–190.

Wright, C. (1998). "Why Frege Did Not Deserve His *Granum Salis*: A Note on the Paradox of 'The Concept Horse' and the Ascription of Bedeutungen to Predicates". *Grazer Philosophische Studien* 55: 239–263.

Yablo, S. (2014). *Aboutness*. Princeton: Princeton University Press.

Peter Hanks
University of Minnesota

Predication and Rule-Following

Abstract: According to the traditional view, propositions are the primary bearers of truth conditions. We latch onto propositions by entertaining them, and then deploy them in various ways in our thoughts and utterances. The truth conditions of our thoughts and utterances are then derived from the truth conditions of propositions. This raises the question of whether we can explain how propositions themselves have truth conditions, which is one form of the problem of the unity of the proposition. Attempts at solving this problem end in a philosophically unsatisfying account of representation in thought and language, in which propositions and our relations to them are all taken as primitive and *sui generis*. In my book *Propositional Content* (Hanks 2015a) I developed a theory that reverses the traditional order of explanation. According to this theory, acts of predication are the primary bearers of truth conditions. Propositions are types of these actions, which derive their truth conditions from their tokens. This puts acts of predication at the heart of the philosophical explanation of representation, content, and truth. Acts of predication are acts in which we sort or categorize objects according to properties. Through their satisfaction conditions, properties provide the correctness conditions for these acts. For example, an act of predicating the property of being blue of an object is correct just in case that object satisfies that property. By providing correctness conditions, properties thus play a crucial role in explaining how acts of predication are capable of being true or false. In this paper I consider two problems for this explanation. The first, which I call the "Stalnaker problem", asks for an explanation of how properties have satisfaction conditions. This is the analog of the unity problem for properties. Solving this problem is relatively straightforward, but it leads to a second, much more difficult problem, which goes to the heart of Wittgenstein's rule-following considerations. These considerations pose serious metaphysical and epistemological problems for a Platonistic conception of rule-following, on which properties provide standards of correctness for acts of predication. I do not currently have an alternative to Platonism. The aim of this paper, then, is to clarify the problems that Wittgenstein's rule-following considerations pose for an act-based theory of predication, and to present some desiderata for a solution.

Keywords: propositions, predication (acts of), properties, rule-following, Ludwig Wittgenstein, objectivity, correctness.

0. Introduction

One of the traditional roles for propositions is to serve as a source of representational properties and truth conditions. This is captured by the idea that propositions are the *primary* bearers of truth conditions. Anything else that has truth conditions, such as a sentence or an assertion or a belief, derives its truth conditions from a proposition. On this traditional account, propositions come first in the order of explanation for truth conditions, and mental and linguistic items come second. Switching images, it is standard to locate propositions in Frege's abstract third realm, above the concrete realm of mental and linguistic states and events. This image endows the traditional view with a top-down explanatory structure, with propositions at the top and mental and linguistic items underneath.

In recent work on the nature of propositional content I have argued for a reversal of the traditional order of explanation (Hanks 2011, 2014, 2015a). On this alternative approach the primary bearers of representation and truth conditions are actions that we perform when we are thinking or speaking. Propositions derive their representational properties and truth conditions from these actions. Representation and truth conditions originate with us, through our thoughts and utterances. Propositions are abstractions from these actions, which we use to classify and individuate mental states and speech acts but which play no foundational, explanatory role.

The most straightforward way to implement this approach is with the type/token distinction. The primary bearers of truth conditions are token mental or spoken actions. Propositions are types of these actions, which inherit their truth conditions from their tokens. In previous writings I used the term 'predication' for these actions. Scott Soames, who independently arrived at a similar view, uses the same terminology.[1] Soames and I thus understand predication as a certain kind of mental or spoken action.

Soames takes predication to be primitive (Soames 2010: 29), but there's no need to resort to that. The act of predication is best understood as a rule-governed act of

---

[1] We use the same terminology, but there is a significant difference between the ways Soames and I understand predication. For Soames, acts of predication are neutral and non-committal, similar to the traditional notion of entertainment. On my view, acts of predication are *not* neutral. A normal, stand-alone act of predicating a property of an object commits the subject to the object's having that property. In this sense, predication is closer to the traditional notions of judgment or assertion. This difference won't matter for this paper, but see Hanks (2015a, ch.1) for an argument that Soames's way of understanding predication is incoherent.

sorting or classifying or categorizing. To predicate a property of an object is to sort that object with other similar objects. This act is rule-governed in the sense that it has a standard of correctness – it can be done correctly or incorrectly. Properties provide these standards of correctness. Suppose you have a pile of marbles in front of you, and you've decided to sort out all the blue marbles. Picking up a marble and putting it into the pile of blue marbles is analogous to predicating the property of being blue of that marble. In this case, the property of being blue provides the standard of correctness for your act of sorting. If the marble is blue then you correctly sorted the marble, and if not then not. Had you been sorting according to the property of being green then the standard of correctness would have been different. A crucial role for properties in acts of predication is thus to fix the correctness conditions for those acts. It is precisely because acts of predication have these correctness conditions that they are capable of being true or false, and hence capable of grounding the truth conditions of propositions. On the act-based conception of propositions, understanding how properties endow acts of predication with correctness conditions is essential for understanding the nature of propositional representation.

In this paper I am going to try to deepen this understanding by raising two problems for this approach to propositions and predication. The first problem, which I will call the *Stalnaker question*, has an easy solution. But this solution raises the specter of a much more difficult problem, which goes to the heart of what is known as Wittgenstein's rule-following considerations. The rule-following considerations do not refute the act-based conception of propositions. They do not show that it is unworkable or incoherent. Rather, they undermine a picture of rule-following that provides a natural stopping point for the explanatory strategy of the act-based conception. The effect of this is to put pressure on the act-based conception to push its explanatory strategy further than it currently stands. I do not know how to extend the act-based account in the required way. The aim of this paper, then, is to pose a problem for the act-based conception that demands further work. If nothing else, I hope to clarify a connection between the debate about the nature of propositions and Wittgenstein's rule-following considerations.

1. The Stalnaker Question

I call it the "Stalnaker" question because it was posed to me by Robert Stalnaker during the question-and-answer period after a talk I gave on propositions.[2]

---

2   At the New Work on Speech Acts Conference, held at Columbia University in September 2013. Jeff King has told me that Stalnaker asked the same question of him on previous occasions.

More background is needed, however, before we can understand the point of his question.

As I argued in the talk, and as I have argued at length elsewhere (Hanks 2015a, ch.2), the main problem for the traditional account of propositions is a version of the problem of the unity of the proposition. The unity problem is in fact a family of related problems, which arise in different ways for different theories of propositions (King 2009). For our purposes, the relevant issue is about explaining or understanding how propositions have truth conditions. The traditional view offers, at least in outline, an explanation of how our beliefs and assertions have truth conditions – they get their truth conditions from propositions. But what can we say, if anything, about how propositions themselves get their truth conditions? Given the top-down explanatory structure of the traditional view, in trying to answer this question we are not allowed to appeal to any of our beliefs or intentions or assertions. More generally, we are not allowed to appeal to any representational mental states or utterances. Propositions must have their truth conditions on their own, without any contribution from us. In the face of this constraint it is natural to look to the internal make-up of a proposition, i.e. its constituents and the way those constituents are related to one another, in trying to explain how it has truth conditions. This is why it makes sense to call this a "unity" problem. Since we cannot appeal to what goes on in thought or speech, in explaining how a proposition is capable of being true or false we have to look to the way in which its constituents are unified into a single, representational whole.

The history of attempts at solving this problem is not encouraging. Both Frege and Russell felt the need to say something about how propositions are unified, and both were dissatisfied with their solutions. Frege appealed to the relation of saturation through which an unsaturated predicate-sense comes to be joined with a complete, saturated name-sense. But as he himself admitted, these notions of being saturated and unsaturated are only metaphors, "only figures of speech" (Frege 1892: 193). It was even worse for Russell. The problems he encountered with the unity of the proposition led him to abandon propositions altogether and adopt in their place his multiple relation theory of judgment.[3] In doing so he gave up the traditional account of propositions.

A thorough examination of Frege's and Russell's attempts at solving the unity problem, and more recent attempts, would take us beyond the scope of this paper.[4] Let's instead consider the initially appealing idea that there is in fact no need to

---

3   See Hanks (2007) for the historical details.
4   See Hanks (2015a, ch. 2) and Hanks (2009).

solve this problem. Maybe the demand for an explanation can be rejected. Perhaps it is a groundfloor, primitive fact that propositions are bearers of truth conditions.[5] It is easy to find oneself inclined to say: propositions are bearers of truth conditions by nature or definition, and so there is no need to try to explain why or how they have truth conditions. On this view, the fact that propositions are the primary bearers of truth conditions is a basic starting point in the philosophical account of mental and linguistic representation.

But taking this line is going to force you into other commitments. First, it is going to be difficult to maintain both that it is primitive that propositions have truth conditions and that they are structured entities with constituents. For, presumably, if they had constituents and structure then we could explain their truth conditions in terms of this structure. But we are now supposing that no such explanation can be given. It looks gratuitous to attribute structure and constituents to propositions if their structures and constituents play no explanatory role. Furthermore, as Trenton Merricks has recently pointed out (Merricks 2015: 199–205), there would be an inexplicable correspondence between the truth conditions of a proposition and its constituents.[6] The proposition that a is F is true if and only if a is F, where this is a primitive, unexplainable fact about this proposition. In addition, the constituents of the proposition are a and the property of being F, or some entities suitably related to a and F, such as Fregean senses. (It would be bizarre to hold that the proposition that a is F has constituents that are completely unrelated to a and F.) But these facts about truth conditions and constituents have nothing to do with each other. They are entirely independent. What an amazing coincidence, then, that the constituents of the proposition are, or are closely related to, exactly the object and property that figure in its truth conditions! If you think that propositions have their truth conditions primitively and you think that they have constituents then you are saddled with this kind of unexplained and arbitrary correspondence.

So, if you reject the demand for an explanation of how propositions have truth conditions then there is considerable pressure on you to regard propositions as simple, unstructured, non-composite entities. That in turn leads to even further commitments. Think about the traditional notion of entertainment – the neutral, cognitive relation by which we come into contact with propositions. How should we understand this relation? It cannot be understood as an operation on the constituents of a proposition, since we are now supposing that propositions

---

5   See Merricks (2015) for a recent example of this view.
6   I am indebted here to conversations with Ben Caplan and Chris Tillman.

do not have constituents. What could it be to entertain one of these simple, *sui generis* propositions? It looks as though entertainment must also be understood as a primitive, unexplainable mental act. The same goes for the act of judgment. Judgment cannot be a mental act performed on the constituents of a proposition, since we have now given up on the idea that propositions have constituents. Nor can judgment be a matter of taking a proposition to be true, since that leads to a regress. *Taking* a proposition to be true is just *judging* it to be true, and so we've analyzed one judgment (judging that p) in terms of another (judging that p is true). Re-applying the analysis, we're left with yet another judgment (judging that [that p is true] is true), and now we're off on a regress. The lesson is that judgment will also have to be regarded as primitive and unexplainable.

This, then, is where the traditional accounts ends up by taking it to be primitive that propositions have truth conditions. There are simple, unstructured, *sui generis* entities that have truth conditions as a matter of unexplainable fact. We latch onto these entities via a primitive mental act of entertainment, and then commit ourselves to them via a primitive mental act of judgment. The resulting judgment then takes on the truth conditions of the proposition.

That is about as empty and unsatisfying a philosophical explanation as one is going to find. It is hardly even worth calling it an explanation. But this is where you end up if you accept the traditional account of propositions and reject the unity problem.

One of the central motivations for the act-based conception of propositions is its ability to avoid this explanatory dead-end. The primary or basic bearers of truth conditions are particular token acts of predication. Propositions are types of these actions, which inherit their truth conditions from their tokens. This is one instance of the more general phenomenon of types inheriting properties from their tokens. Consider the Union Jack, the flag of the United Kingdom. Tokens of the Union Jack are rectangular, striped, and partly red and partly blue. But so is the *type* of flag. It is perfectly natural and commonsensical to say that the Union Jack (the type) is rectangular, striped, partly red and partly blue. But how can this abstract type have these properties? It has them in a secondary or derivative sense. The primary bearers of these properties are tokens of the Union Jack. The fact that the Union Jack, the type, is rectangular, striped, etc. is constituted by the fact that any possible token of this type is rectangular, striped, etc. The type derives these properties from its tokens in the sense that its having these properties is nothing more than the possession of these properties by any of its possible tokens. The same holds for predication and truth conditions. A type of act of predication has truth conditions insofar as any possible token of that type has those truth conditions. This is what I mean when I say that types of acts of predication (propositions) *inherit* their truth conditions from their tokens.

We are now ready to understand Stalnaker's question. His question was: why do properties have satisfaction conditions? This is the analog for properties of the unity problem for propositions. Remember that the relevant form of the unity problem for propositions was about explaining how propositions have truth conditions. The analog for properties is a question about satisfaction conditions. Properties are not true or false, but they are *satisfied* or *unsatisfied* by objects. The property of being blue is satisfied by an object x just in case x is blue. Stalnaker's question asks for an explanation of why or how this property has these satisfaction conditions.

Now, the overwhelmingly natural response to this question is to reject the demand for such an explanation. Once again it is easy to feel inclined to say: it is a primitive, groundfloor fact about properties that they have satisfaction conditions. That is just what properties are, by nature or definition – the sorts of things that can be satisfied by objects. I'm fairly sure that Stalnaker thought this was the right response, and it surely is. The point of his question was *not* to demand an explanation of how properties have satisfaction conditions, but rather to challenge the idea that we have to give a corresponding explanation for propositions.[7] If there's no need to explain why properties have satisfaction conditions then, similarly, there ought to be no need to explain why propositions have truth conditions. But as we've seen, there are good reasons for not rejecting the explanatory demand in the case of propositions. Rejecting the unity problem for propositions leads to a vacuous account of how our thoughts and utterances acquire their truth conditions.

Fortunately, there is a crucial disanalogy between properties and propositions that breaks the parity between the two cases. Propositions are *representational* and *intentional*, whereas properties are neither. Truth and falsity are representational properties. A proposition is true when it represents the world to be a certain way and the world is that way.[8] The proposition that *a* is F represents *a* as being F. The

---

7   Thanks to Gary Ostertag who helped me, through personal communication, to see the point of Stalnaker's question. See Ostertag (2013) for his discussion of the problem.
8   Some recent accounts of propositions deny that propositions are representational, e.g. Speaks (2014a, 2014b) and Richard (2013). These philosophers hold that propositions have truth conditions but deny that they represent anything. As Soames has pointed out, this commits them to denying the platitude that "truth is a kind of accuracy in representation … a proposition is *true* when it represents things as they really are" (Soames 2014: 167). See Speaks (2014b) for a response, and Hanks (2015b) for discussion.

proposition *says* that *a* is F. This is what it makes it possible for the proposition to be true or false. Similarly, propositions are intentional. The proposition that *a* is F is *about a*; it is concerned with or directed at *a*. These features of propositions, being representational and intentional, call out for philosophical explanation. They are not something that we can reasonably take as primitive and then use to explain the representationality and intentionality of thought and speech. Properties, by contrast, are neither representational nor intentional. A property does not represent the objects in its extension. It does not say anything about those objects, and it is not about them or directed at them. Satisfaction is not a representational or intentional relation. Satisfaction conditions therefore do not pose the same kind of philosophical problem that truth conditions pose.

This means that we can reject the demand for explanation contained in Stalnaker's question. We can safely take it to be primitive that properties have satisfaction conditions, without thereby undermining the need for an explanation of how propositions have truth conditions.

2. Platonism about Rule-Following

But now another problem emerges, which is not so easily solved. The satisfaction conditions of properties play a crucial role in the act-based conception's explanation of how propositions have truth conditions. That explanation, recall, starts with the idea that particular token acts of predication are the basic or primary bearers of truth conditions. Why do these token acts of have truth conditions? Because they are acts of rule-governed sorting or classifying or categorizing. To predicate the property of being F of *a* is to sort *a* with other objects, where the property of being F provides the standard of correctness for this act of sorting. When you sort *a* according to this property you can get it right or wrong – your act can be correct or incorrect. What condition must be met in order to correctly sort *a* according to this property? Well, it has to be that *a* has that property, that *a* is F. Being F is the condition that *a* must meet if this act of sorting is to be correct. This is the sense in which the property provides the standard of correctness for the act of sorting. Furthermore, the act of predication has truth conditions because it is subject to this standard of correctness. The truth or falsity of an act of predication is just a matter of whether the object is correctly or incorrectly sorted or classified or categorized. But *a* is correctly sorted with the other F's just in case *a satisfies* F. The satisfaction conditions for the property F *are* the correctness conditions for acts of predicating F of objects. The possession of truth conditions by acts of predication thus depends crucially and directly on the possession of satisfaction conditions by properties.

Following our answer to the Stalnaker question, we are now taking it to be primitive that properties have satisfaction conditions. We are viewing properties as mind and language independent entities that have satisfaction conditions by their natures or essences. We are then using these properties to explain how token acts of predication have truth conditions. We latch onto them and apply them to objects in acts of predication. Through their satisfaction conditions properties provide the standards of correctness for these acts of predication. Any such act is correct (true) or incorrect (false) depending on whether the object satisfies the property.

This picture of how our acts of predication come to be correct or incorrect has variously been called "classical realism" about meaning (Kripke 1982; Wilson 1994, 1998, 2006) and the "classical account of extension" (Wilson 1982), the "rules-as-rails" conception of meaning (Wright 1989a, 1989b), and "Platonism" about meaning (Wright 2001; Zalabardo 2003). It is the picture of rule-following that Wittgenstein seems to have in mind when he discusses the "superlative fact" which you grasp "in a flash" when you grasp the meaning of a word (*PI* 191–2), and in his image of "rails invisibly laid to infinity" (*PI* 218), which lay down ahead of time the conditions that have to be met in order for any possible application of a word to be correct. George Wilson sums up the picture as follows:

> If a speaker means something by a general term 'Φ', then the speaker has adopted a rule that specifies the standards of correctness for 'Φ' as she proposes to use it. The rule, we may suppose, has for her the form: 'Φ' (as I shall use it) is to be ascribed to an object *o* just in case *o* satisfies *those conditions*, where the conditions are given by some property or properties that the speaker has suitably in mind. These conditions are comprised of properties that exist independently of language and are exemplified (when they are), independently of our ability to ratify the relevant facts. (Wilson 2006: 153)

The point in Wilson's last sentence is worth emphasizing. On this Platonist account (as I will call it), the property associated with a word provides an *objective* standard of correctness for applications of that word. It is an objective matter whether an object satisfies a property. Whether it is correct or incorrect to apply a predicate to an object is settled independently of any of our views about whether it is correct or our ability to determine if it is correct. It is possible for all of us, individually and collectively, to get it wrong about whether an object satisfies a property. In fact, the Platonist picture can be viewed as a natural outcome of this demand for objectivity. By locating the source of correctness conditions in mind and language independent entities, which have their satisfaction conditions primitively and completely independently of us, Platonism offers a very pure account of how acts of predication are subject to objective standards of evaluation.

There is also an important epistemological dimension to the Platonist picture. Properties not only provide the correctness conditions for acts of predication, they must also serve as guides for our classificatory activities. Suppose you are trying to decide whether to sort a marble into the pile with the other blue marbles. The property of being blue provides the condition that the marble must meet in order to be correctly sorted into that pile. You have to decide, then, whether the marble meets this condition. If your decision is disconnected from this condition – if the condition plays no role in your deliberations – then whatever decision you reach will be unjustified (cf. Zalabardo 1997). It's not enough just to examine the marble. Your apprehension of the condition determined by the property of being blue must play a role in this examination. In examining the marble you have to check to see whether it has the feature determined by the property of being blue that something must have in order to satisfy it. If your decision about how to sort the marble is not informed by this property then any decision you make will be no better than a guess. It would be like saying 'It's raining' without paying any heed to the conditions that have to be met for it to be raining. For our acts of predication to be justified, then, it has to be that we engage with the properties that provide their standards of correctness. Given that properties provide these standards of correctness, for our acts of predication to be justified we must use them as guides in our predicative activities.

As we will see, these two roles for properties on the Platonist account, i.e. their provision of objective standards of correctness, and their role as guides for acts of predication, are the sources of the problems for Platonism. The demand for objectivity leads the Platonist to view properties as transcendent entities, utterly removed and independent of minds and languages – a view which is reinforced by the idea that properties are primitive bearers of satisfaction conditions. This leads to metaphysical problems about making sense of how we are capable of latching onto these properties in order to fix correctness conditions for our acts of predication. This same transcendence is the source of epistemological problems about understanding how we could possibly be guided by properties in our acts of predication. But we are getting ahead of ourselves. Before discussing these problems I want to forestall an objection.

The Platonist view as discussed in the rule-following literature is an account of the meanings of words – in particular, the meanings of predicates. The view is that meaning something by a predicate is a matter of associating a property with that predicate, where the property provides the standard of correctness for applications of that predicate to objects. Because of the emphasis on words and meaning, one might be tempted to think that the Platonist account can be decoupled from the theory of predication. Some acts of predication are linguistic and involve words, but not all of them do. Acts of predication can occur in thought. When I judge that *a* is F I mentally predicate F of *a*, and no words have to be involved in this act of predication. Acts of predication can also be perceptual or behavioral. When sorting the marbles I am performing acts of predication that involve looking at the marbles and physically putting them into different piles. I can do all of that without uttering any words, either outwardly or in my head. These are non-linguistic, mental or perceptual or behavioral forms of predication. Given that the Platonist picture is about the meanings of words, it is tempting to say that it is irrelevant to understanding how these non-linguistic acts of predication come to have correctness and truth conditions. Furthermore, if mental or perceptual or behavioral acts of predication are the explanatorily basic bearers of representation and intentionality, then we could make these kinds of acts our sole focus and relegate linguistic acts of predication to secondary status. The act-based conception of propositions would then become the view that propositions are types of *cognitive* acts of predication (where 'cognitive' covers non-linguistic judgment, perception, and intentional behavior). Insofar as these cognitive acts are independent of language, considerations about the meanings of linguistic expressions are irrelevant.

There are two things to say in response. The first is that any non-linguistic, cognitive act of predication seems to involve or require a mental act of predication, and there is a widespread view in cognitive science according to which mental acts of predication employ linguistic vehicles, although not necessarily expressions in a public language (Fodor 1975, 2008). If there is a Language of Thought, then a mental act of predicating F of *a* consists in combining a mentalese expression for *a* with a mentalese expression for F. The Platonist picture can then be reinstated through its account of the meanings of mentalese predicates. These predicates are meaningful because they are associated, perhaps by naturalistic relations, with properties that provide standards of correctness for their application.

But, secondly, and more importantly, there is no need to adopt the Language of Thought hypothesis in order to tie the Platonist account to non-linguistic acts of predication. Any act of predication is an act of sorting an object according to a rule. Performing an act of predication therefore requires determining a rule for this act of sorting. That is, it requires determining correctness conditions. According to the Platonist account, properties are the source of these correctness conditions. We fix upon a rule for an act of sorting by fixing on a property. In a linguistic act of predication this act of fixing on a property is mediated by a linguistic expression. First the property is associated with the expression, and then applications of this expression take on the correctness conditions given by the property. In non-linguistic predication no linguistic vehicle is involved. In these cases, the subject gives her act of predication correctness conditions in a direct and unmediated way by singling out a property. This is still the Platonist picture, just without linguistic intermediaries. Following Wilson's formulation, we can state this general form of Platonism as follows:

> If a subject engages in an act of predication then the subject has adopted a rule that specifies the standards of correctness for that act of predication. The rule, we may suppose, has for her the form: this act of predication, targeted at object *o*, is correct just in case *o* satisfies *those conditions*, where the conditions are given by some property or properties that the subject has suitably in mind. These conditions are comprised of properties that exist independently of language and are exemplified (when they are), independently of our ability to ratify the relevant facts.

This generalized version of Platonism applies to any act of predication, whether it involves words or not. In the remainder of this paper by 'Platonism' I will mean this more general picture.

3. The Metaphysical and Epistemological Problems for Platonism

As I mentioned earlier, there are two sorts of problems for the Platonist view, one metaphysical and the other epistemological. Let's start with the metaphysical problems.

Although not presented in quite these terms, the metaphysical difficulties for Platonism have their clearest and best known expression in Kripke's book on Wittgenstein (Kripke 1982).[9] What facts about me, my mental states or my behavior, determine one property as opposed to another as the standard of correctness for an act of predication? For example, why does the property of being blue give the correctness condition for my act of sorting the marble, as opposed to the property of being bleen?[10] Kripke's Wittgenstein finally reaches the conclusion that there is nothing about me that determines one property as opposed to infinitely many

---

9   Kripke presents his Wittgensteinian skeptical arguments as a general challenge to the existence of semantic facts, e.g. the fact that Jones means blue by 'blue'. We can read them, however, not as arguments for a global skepticism about meaning, but as a challenge to one particular picture of what meaning facts consist in, namely the Platonist picture. See Wilson (1994, 1998, 2006) for a reading of Kripke along these lines.
10 Something is bleen iff either it is observed before some future time *t* and blue or it is green. 'Bleen' is a close cousin of 'grue', both of which are due to Goodman (1983).

others. Nothing in my past behavior, or any instructions I might give myself, or mental images I might entertain determines one property out of infinitely many deviant alternatives. Furthermore, although this is more of a vexed issue, my dispositions for sorting the marbles are not sufficient to determine blue instead of some bleen-like property.[11] After considering these and other proposals, Kripke concludes on Wittgenstein's behalf that there are no facts about me that succeed in singling out any property to be the standard of correctness for my act of predication. I won't rehearse the battery of detailed arguments that Kripke provides for this conclusion, and I can't review the large literature that responds to these arguments. It is enough for my purposes to point out that the theory of predication we have so far been considering faces the skeptical challenge that Kripke presents in his book on Wittgenstein. That theory of predication assumes that we identify properties and use them to impose standards of correctness on our acts of predication. Kripke's skeptical challenge demands that we explain exactly how we single out properties to serve as these standards of correctness.

Now, it might be thought that Kripke overlooked an easy and obvious answer to his skeptical challenge (cf. Zalabardo 2003). Maybe we have a quasi-perceptual faculty for grasping properties. Russell thought that "it is obvious ... that we are acquainted with such universals as white, red, black, sweet, sour, loud, hard, etc." (Russell 1912: 101). But this proposal has to be developed carefully in order to avoid Kripke's skeptical arguments. On Russell's account, we become acquainted with universals by a process of abstraction from particulars. First we are acquainted with the various blue marbles (or, in Russell's case, with the corresponding sense data). By a process of abstraction on these marbles (or sense data) we become acquainted with the property of blue found in each. But what determines that this process of abstraction determines the property of being blue as opposed to the property of being bleen? How does abstraction lead us to the blue-ness of the marbles instead of their bleen-ness? Perhaps it is a matter of attending to the blueness of the marbles instead of their bleen-ness. But how do you do that? Any blue

---

11 See Kripke (1982: 22–37) for the original discussion of the dispositionalist response, and Boghossian (1989) for an overview of the problems and prospects for dispositionalism in the wake of Kripke's discussion. One problem for dispositionalism that hasn't received much attention (although see Boghossian 2015) is that it gets the order of explanation wrong. Intuitively, you are disposed to sort the blue marbles into the blue pile *because* you are sorting according to the property of being blue. According to the dispositionalist response, however, the situation is exactly the reverse: you are sorting according to the property of being blue *because* you are disposed to sort the blue marbles into the separate pile.

marble we have observed before *t* is also bleen, and its blue color is also its bleen color. So how can you attend to its blue-ness without also attending to its bleenness? These questions land us back into the realm of Kripkean skeptical considerations that this move was supposed to avoid. Better, then, not to follow Russell in giving an abstractionist account of how we grasp properties. In fact, better not to give *any* substantive or reductive account of how we become acquainted with properties, since any such account will be exposed to the Kripke-Wittgenstein skeptical challenge. Any substantive account of how I grasp a property will supply conditions that can be reinterpreted in such a way that they count as acquaintance with a deviant property. Grasping a property looks like it has to be understood as a *sui generis*, primitive form of cognitive contact between a subject and a property. The properties are out there, and through a primitive mental act of grasping we single them out for our acts of predication.

If we were dissatisfied with the earlier account of propositional representation, on which propositions are primitive bearers of truth conditions that we come into contact with through primitive relations of entertainment and judgment, then I think we should be equally dissatisfied with this appeal to a primitive mental act of grasping properties. The two views look disturbingly similar. The first view posits primitive bearers of truth conditions and primitive relations of entertainment and judgment. The account of predication we are now considering posits primitive bearers of satisfaction conditions and a primitive relation of grasping or acquaintance. We have made some progress, since we are not taking representation or intentionality to be primitive, but the explanation of how acts of predication acquire their correctness conditions still looks empty and unsatisfying.

But let's suppose that these metaphysical problems can be solved and that a satisfying explanation can be given of how a subject singles out a property and thereby imposes standards of correctness on her act of predication. This leaves untouched the epistemological problems for Platonism. The clearest and most forceful articulation of these problems can be found in Crispin Wright's work on the rule-following considerations (collected in Wright 2001). Recall the epistemological role for properties in the Platonist account. In order for an act of predication to be justified, the subject has to be guided by whatever determines the standard of correctness for that act. Since, on the Platonist view, properties determine these standards of correctness, it follows that properties must serve as guides for our acts of predication. The epistemological problems have to do with making sense of how we could be so guided.

These problems can be put in the form of dilemma. Either I am guided by a property through a direct, intuitive grasp of what the property requires, or I am guided indirectly, through concrete instructions or a formula or template. The first horn just adds another bit of primitive machinery to the account of predication we have just been considering. There are mind and language independent primitive bearers of satisfaction conditions, which we latch onto through a primitive relation of grasping or acquaintance, and which guide us in our acts of predication through primitive acts of intuition. This is exactly the sort of philosophically unsatisfying explanation that we are trying to avoid by adopting the act-based conception of propositions. Perhaps we could streamline the view by collapsing the relations of grasping and intuition into one. That is, perhaps we single out properties and intuit their requirements through a single, primitive relation of grasping/intuition. That simplifies the view, but it doesn't dispel any of its mysteriousness. Furthermore, as Wright has emphasized, respecting the objectivity of the standard provided by a property means that we have to recognize the possibility of *wrongly* intuiting the standards of correctness determined by a property (Wright 1989a: 161). It has to possible for me to know all the relevant facts about the object in question and still get it wrong about whether it satisfies a property because I have incorrectly intuited the conditions determined by that property for an object to fall under it. What would it be to incorrectly intuit the requirements of a property in this way? There is nothing to say about this, since we have bundled away the notions of grasping and intuiting into a black box. As Wright puts it, "nothing can be done by way of *filling out* the thought that, in the most primitive cases of rule-following, when everything seems immediate and beyond further account, we nevertheless track a set of independent requirements" (Wright 1989a: 161).

The second horn, on which I am guided by a property indirectly through instructions of some kind, either leads back to the first horn or it generates a regress. Suppose the instructions I give myself for sorting the marbles consists of a blue tile, which I use as a guide in sorting the marbles. I make decisions about how to sort the marbles by comparing them with the blue tile. The problem is that the tile *on its own* cannot guide me in sorting the marbles. I need some further instructions or interpretation about how to use it. When I compare a marble with the tile, what am I looking for? How do I use the tile to decide whether the marble should be sorted into the separate pile? The obvious answer is that I am checking to see whether the marble is *similar in color* to the tile. That is the additional instruction I am using alongside the blue tile. It is the blue tile, along with the instruction to check for similarity in color, which serves as a guide in my acts of predication. But now the whole problem re-occurs, since I am in effect sorting the marbles according to the property *similar in color to this tile*. How can I be guided by this property in sorting the marbles? Our original dilemma now arises all over again.

Either I directly intuit this property, or I have some instructions that indirectly inform me about its requirements. The first option puts us back on the first horn of the dilemma, and the second starts off a regress.

It might be tempting to think that these problems can be avoided if the instructions are mental items, rather than physical or material objects of some kind. Maybe a blue tile on its own cannot guide me in sorting the marbles, but a mental image of blueness could. But this gets us nowhere. A mental image is no more capable of fixing its own interpretation than a concrete blue tile. As Wittgenstein points out in the Blue Book, the temptation to think otherwise is based on the tendency to endow the mental with an "occult character", which allows it to "bring about effects which no material mechanism could" (Wittgenstein 1960: 3–5). He recommends the following as a way to resist this tendency:

> There is a way of avoiding at least partly the occult appearance of the processes of thinking, and it is to replace in these processes any working of the imagination by acts of looking at real objects. Thus it may seem essential that, at least in certain cases, when I hear the word "red" with understanding, a red image should be before the mind's eye. But why should I not substitute seeing a red bit of paper for imagining a red patch? The visual image will only be the more vivid. (Wittgenstein 1960: 4)

Take the blue mental image that occurs to you while sorting the marbles and replace it with a corresponding blue piece of paper. Or, somewhat more fancifully, imagine that you could take the mental image out of your mind and convert it into a concrete image. If the latter is insufficient for guiding your sorting behavior then so is the former. The idea that the mental image is capable of something that the material image is not is a confusion borne of giving the mind magical powers.

We have seen, then, that there are serious metaphysical and epistemological problems for the Platonist account of how acts of predication acquire their standards of correctness. The metaphysical problems arise when we try to give an account of how a subject singles out a property that will provide this standard of correctness. The epistemological problems arise when we try to make sense of how a subject is guided by this property in making decisions about which acts of predication to perform. The difficulties these problems present tend to lead us to give up the attempt at explaining any of this, and to treat the relations of grasping and intuiting properties as primitive. The resulting picture of how acts of predication acquire their correctness conditions, which posits primitive bearers of satisfaction conditions, and primitive relations of grasping and intuiting, is not much of an advance over the traditional account of propositions, which, in order to avoid the unity problem, has to posit primitive bearers of truth conditions and primitive relations of entertainment and judgment. To make progress we are going to have to abandon Platonism and look elsewhere for an account of how acts of predication are capable of being correct or incorrect.

4. Objectivity, Justification, and Generality

As I mentioned in the introduction, I do not have an alternative to Platonism. Such an alternative would extend the explanatory strategy of the act-based conception of propositions by providing a non-Platonistic account of how acts of predication acquire their correctness conditions. I do not currently know how to do this. Or rather, I do not know how to provide an alternative to Platonism that can meet the twin demands of preserving objectivity in the standards of correctness for acts of predication and our justification for those acts.

Consider, for example, a *communitarian* account of correctness conditions, on which the correctness of an act of predication is a matter of whether other people would evaluate it as correct.[12] On this view, when I sort a marble into the pile with the other blue marbles, this act of sorting is correct just in case some sufficient number of people in my community would agree that the marble belongs in that pile. Correctness, then, consists in agreement with other people's opinions or assessment of whether an act of predication is correct.

Properties, understood as primitive bearers of satisfaction conditions, play no role in this version of communitarianism. The standard of correctness for an act of predication does not come from a property but rather from what other people do or would say about whether the act is correct. There is another form of communitarianism on which properties still play this role. This is, in effect, a communal form of Platonism. On this view, the facts that single out properties to serve as standards of correctness are facts about whole communities, not individuals taken one by one. This view is easiest to make sense of when taken as an account of the meanings of predicates. The predicate 'blue' means the property of being blue because of community wide facts about how people use the word 'blue'. My meaning blue by 'blue', and hence being subject to this standard of correctness when I apply the word 'blue', is a matter of my being a member of this linguistic community. It

---

12 This communitarian account is not the Skeptical Solution that Kripke offers on behalf of Wittgenstein in Kripke (1982), although it bears affinities to the Skeptical Solution. The Skeptical Solution supplies assertability conditions for semantic claims of the form 'Jones means blue by 'blue'' which depend on agreement between the ways in which Jones uses the word 'blue' and the way others in his community are disposed to use this word. The communitarian account, by contrast, supplies correctness conditions for Jones's uses the word 'blue' that depend on this agreement.

is harder to make sense of this view when applied to non-linguistic acts of predication. What facts about the community could figure in determining the property that I am going to use for sorting the pile of marbles? It is not at all clear what to say about this. The Platonistic communitarian account is also exposed to the same metaphysical and epistemological problems that its individualistic version faces. What facts about the community as a whole single out one property as opposed to infinitely many other non-standard alternatives? Furthermore, given that a property has been determined, how can individuals in the community be guided by it in their acts of predication? The shift from an individual to a community doesn't seem to offer much hope for progress on these questions.

The communitarian account we are now considering is a genuine alternative to Platonism. It dispenses with properties altogether and locates the source of correctness directly in the actual and potential verdicts that people in the community would give if queried about the appropriateness of an act of predication. The obvious problem for this form of communitarianism is the loss of objectivity that it entails. McDowell articulates this conception of objectivity in a well-known passage:

> The idea at risk is the idea of things being thus and so anyway, whether or not we choose to investigate the matter in question, and whatever the outcome of any such investigation. That idea requires the conception of how things could correctly be said to be anyway – whatever, if anything, we in fact go on to say about the matter ... (McDowell 1986, 46)

The example I've been using so far, of sorting out the blue marbles, may not be the best one for illustrating this point, since you might think that it is not an entirely objective matter whether something is blue (e.g. because you think that colors are secondary qualities). So suppose that the marbles are made of different materials – glass, plastic, metal, and agate – and I am sorting the glass marbles into a separate pile. Whether a marble is made of glass is a thoroughly objective matter, which is independent of what any or all of us might think or say about it. We might all collectively get it wrong about whether a marble is made of glass, even after careful and thorough examination. This is the conception of objectivity that is put under threat by the communitarian account. According to this account, it is correct to predicate being made of glass of a marble just in case a sufficient number of us are disposed to accept this act of predication as correct. It is a short step from this to the absurd sounding conclusion that the marble is made of glass just in case a sufficient number of us are inclined to say that it is made of glass.

The communitarian account must also answer the epistemological questions that arose for Platonism. If our acts of predication are to be justified, it must be that we are somehow guided by the facts that determine the standards of correctness for those acts. On the communitarian account, these are facts about the opinions of other people in the community. It follows that in order for an act of predication to be justified, I have to check to see whether a sufficient number of other people in my community would agree to that act of predication. We rarely do that when performing acts of predication. We blithely go about categorizing things in our everyday acts of judgment and assertion without paying any attention to whether there is a consensus behind those acts. Given the communitarian account, this leads to wide-ranging skepticism. If correctness is a matter of communal agreement, and we are not generally guided by considerations about such agreement when performing acts of predication, then it turns out that these acts of predication are unjustified. The communitarian account manages to combine an anti-realist abandonment of objectivity with general epistemological skepticism. It is a genuine alternative to Platonism, but it is not something we should want anything to do with.

A viable alternative to Platonism, then, will preserve objectivity in the correctness conditions for acts of predication while not leading to wholesale skepticism about the justification for those acts. When I say that I do not have an alternative to Platonism what I mean is that I don't have an alternative that meets these desiderata.

I would like to close by considering a third desideratum, whose status is less firm than those of objectivity and the avoidance of skepticism. One of the appealing features of Platonism is its generality – the way in which it provides a single, universal account of how acts of predication acquire their correctness conditions. Although the properties involved vary from one act of predication to another, the general story about the role of properties in endowing acts of predication with correctness conditions stays fixed. Despite its other faults, the communitarian view also exhibits this level of generality. The communitarian account provides a short, neat statement of what it takes for an act of predication to be correct – it has to be accepted, or would be accepted, by a sufficient number of people in the relevant community. The third desideratum for an alternative to Platonism, then, is that it tells this sort of general story about the source of correctness conditions for acts of predication.

Abandoning this desideratum would mean abandoning the search for a *theory* about how acts of predication acquire their correctness conditions. Perhaps it is possible to give accounts of correctness conditions on a case-by-case basis. When I'm sorting out the blue marbles there is a feature in the marbles themselves that I am looking for – a feature that I can point to and illustrate by holding up one of the marbles. Correctness in this case, then, is a matter of whether a marble exhibits this feature. If I am sorting out the marbles that are made of glass I'll have to tell a different story about what it takes for an act of predication to be correct. In that case, the story will refer to the material constitution of the marbles. What we have to resist is the thought that these two accounts can be unified into a more general theory about where the correctness conditions for acts of predication come from.

The idea that we have to give up on theorizing and content ourselves with descriptions of particular cases is a familiar theme in the *Investigations*:

> And we may not advance any kind of theory. There must not be anything hypothetical in our considerations. We must do away with all *explanation*, and description alone must take its place. And this description gets its light, that is to say its purpose, from the philosophical problems. These are, of course, not empirical problems; they are solved, rather, by looking into the workings of our language, and that in such a way as to make us recognize these workings; *in despite of* an urge to misunderstand them. (*PI* 109)

These and similar remarks are usually read as expressions of quietism. In the present context, quietism takes the form of Wittgenstein's refusal to offer an alternative to the Platonist picture that he undermines with the rule-following considerations. This refusal is usually met with dissatisfaction or frustration (e.g. Wright 1989: 169). But I think it helps to see this refusal, not as simple, stubborn unwillingness to say anything constructive, but rather as a rejection of the philosophical urge for generality. There are plenty of things we can say about how our acts of predication acquire their correctness conditions, but these explanations will be local and piecemeal. The problems may come when we try to unify everything together into a single global account.

5. Conclusion

The act-based conception of propositions locates the source of representation and truth conditions in acts of predication. This constitutes a reversal of the traditional order of explanation. Instead of thoughts and utterances inheriting truth conditions from propositions, propositions inherit their truth conditions from the acts of predication we perform in thinking and speaking about the world. Making good on this reversal, however, requires giving an explanation of how acts of predication acquire their truth conditions. It is natural for this explanation to take the form of a Platonist account of rule-following, on which our acts of predication derive their correctness conditions from mind and language independent properties. But this Platonist account is exposed to the powerful metaphysical and epistemological problems that have emerged in the literature on Wittgenstein's rule-following considerations. The aim of this paper was to clarify these problems for the act-based conception of propositions and to explore, in a very preliminary form, avenues for a solution.

References

Boghossian, P. (1989). "The Rule-Following Considerations.". *Mind* 98: 507–549. Reprinted in A. Miller and C. Wright (2002), 141–187.

Boghossian, P. (2015). "Is (Determinate) Meaning a Naturalistic Phenomenon?". In: S. Gross, N. Tebben and M. Williams (eds.), *Meaning Without Representation: Essays on Truth, Expressivism, Normativity,and Naturalism*, Oxford: Oxford University Press, 331–358.

Fodor, J. (1975). *The Language of Thought*. Cambridge, MA: Harvard University Press.

Fodor, J. (2008). *LOT 2: The Language of Thought Revisited*. New York: Oxford University Press.

Frege, G. (1892). "On Concept and Object." In: M. Beaney (ed.), *The Frege Reader*, Oxford: Blackwell, 181–193.

Goodman, N. (1983). *Fact, Fiction, and Forecast*. Cambridge, MA: Harvard University Press.

Hanks, P. (2007). "How Wittgenstein Defeated Russell's Multiple Relation Theory of Judgment". *Synthese* 154: 121–146.

Hanks, P. (2009). "Recent Work on Propositions". *Philosophy Compass* 4/3: 469–486.

Hanks, P. (2011). "Structured Propositions as Types". *Mind* 120:11–52.

Hanks, P. (2014). "What are the Primary Bearers of Truth?". *Canadian Journal of Philosophy, Special Issue: Essays on the Nature of Propositions*, D. Hunter and G. Rattan (eds.), 43: 558–574.

Hanks, P. (2015a). *Propositional Content*. Oxford: Oxford University Press.

Hanks, P. (2015b). "Review of *New Thinking About Propositions* by Jeffrey C. King, Scott Soames, and Jeff Speaks". *Notre Dame Philosophical Reviews,* http://ndpr.nd.edu/news/50321-new-thinking-about-propositions/.

King, J. C. (2009). "Questions of Unity," *Proceedings of the Aristotelian Society* 109: 257–277.

King, J. C., Soames, S. and Speaks, J. (2014). *New Thinking About Propositions*. Oxford: Oxford University Press.

Kripke, S. (1982). *Wittgenstein on Rules and Private Language*, Cambridge, MA: Harvard University Press.

McDowell, J. (1984). "Wittgenstein on Following a Rule". *Synthese* 58: 325–363. Reprinted in A. Miller and C. Wright (2001), 45–80.

Merricks, T. (2015). *Propositions*. Oxford: Oxford University Press.

Miller, A. and Wright, C. (eds.). 2002. *Rule-Following and Meaning*. Montreal: McGill-Queen's University Press.

Ostertag, G. (2013). "Two Aspects of Propositional Unity". *Canadian Journal of Philosophy, Special Issue: Essays on the Nature of Propositions*, D. Hunter and G. Rattan (eds.), 43: 518–533.

Richard, M. (2013). "What are Propositions?". *Canadian Journal of Philosophy, Special Issue: Essays on the Nature of Propositions*, D. Hunter and G. Rattan (eds.), 43: 702–740.

Russell, B. (1912). *The Problems of Philosophy*. Oxford: Oxford University Press.

Soames, S. (2010). *What is Meaning?* Princeton: Princeton University Press.

Soames, S. (2014). "Propositions vs Properties and Facts". In: J. C. King, S. Soames and J. Speaks (2014), 166–181.

Speaks, J. (2014a). "Propositions as Properties of Everything or Nothing". In: J. C. King, S. Soames and J. Speaks (2014), 71–90.

Speaks, J. (2014b). "Representational Entities and Representational Acts". In: J. C. King, S. Soames and J. Speaks (2014), 147–165.

Wilson, G. (1994). "Kripke on Wittgenstein on Normativity". *Midwest Studies in Philosophy*, 19: 366–390. Reprinted in A. Miller and C. Wright (2002), 234–259.

Wilson, G. 1998. "Semantic Realism and Kripke's Wittgenstein". *Philosophy and Phenomenological Research* 83: 99–122.

Wilson, G. (2006). "Rule-Following, Meaning, and Normativity". In: E. Lepore and B. Smith (eds.), *The Oxford Handbook of Philosophy of Language*, Oxford: Oxford University Press, 151–74.

Wilson, M. (1982). "Predicate Meets Property". *The Philosophical Review* 91: 549–589.

Wright, C. (1989a). "Critical Notice of Colin McGinn's *Wittgenstein on Meaning*". *Mind* 98: 289–305. Partially reprinted in Wright (2001), 143–169.

Wright, C. (1989b). "Wittgenstein's Rule-Following Considerations and the Central Project of Theoretical Linguistics". In: A. George (ed.), *Reflections on Chomsky*, Oxford: Blackwell, 233–264. Reprinted in C. Wright (2001), 70–213.

Wright, C. (2001). *Rails to Infinity*. Cambridge, MA: Harvard University Press.

Wittgenstein, L. (1953). *Philosophical Investigations*, Oxford: Blackwell.

Wittgenstein, L. (1960). *The Blue and Brown Books*. Second edition, New York: Harper.

Zalabardo, J. (1997). "Kripke's Normativity Argument". *Canadian Journal of Philosophy* 27: 467–488.

Zalabardo, J. (2003). "Wittgenstein on Accord". *Pacific Philosophical Quarterly* 84: 311–329.

Bjørn Jespersen
VSB-Technical University of Ostrava

Is Predication an Act or an Operation?

Abstract: This paper contrasts a logical or objectivist conception of predication with a pragmatic or subjectivist conception. The former conception identifies predication with an objective operation that applies a property to an individual. This conception is suitable as an account of how to provide propositional unity. The latter conception identifies predication with an act that a speaker or thinker carries out when asserting or thinking that an object has a property. According to contemporary act-theoretic theories, this conception accounts for propositional unity. I argue that the two conceptions of predication are complementary rather than mutually exclusive, but also that the logical conception enjoys conceptual priority.

Keywords: predication, proposition, procedural semantics, unity, logic, semantics

> It is folly to look for an account of propositional unity;
> this is just a primitive fact to be regarded with Wordsworthian natural piety.
> Manuel García-Carpintero (2010: 288)

0. Introduction

The question I raise is how predication might undergird propositional unity. Predication may be conceived as an *act* that speakers or thinkers execute when predicating a property of an individual, or as an *operation* that applies to a pair consisting of a property and an individual. I argue that the two approaches to predication are complementary rather than mutually exclusive; but also that the operation-theoretic approach enjoys conceptual priority over its act-theoretic counterpart. When agents are claimed to predicate properties of individuals, the question arises what the logical operation is that underlies their acts of predication.

I go on to argue that when predication is construed as functional application then we are well-positioned to account for the systematic interaction between a property and an individual within the structured proposition that the individual has the property. This description of predication will be qualified below so as to fit into a procedural semantics. A procedural semantics for sentences identifies their meaning with a procedure that yields a truth-condition, rather than with the resulting truth-condition. A procedural semantics compares to a flow chart of logical processes. Procedures are not unlike the linguist's trees, with the important difference that trees are, logically speaking, nothing but set-theoretic sequences, while procedures are mereological objects, i.e. wholes with parts.

Finally I argue that structured propositions must be identified with procedures that contain sub-procedures as their parts. The atomic proposition that a is an F is identified with a complex procedure some parts of which identify a and F and prescribe the predication of F of a. The adjacent type theory makes it clear that the result of the application is a truth-value, and that the truth-value is abstracted over in order to yield a truth-condition. Four different sorts of procedures are involved in the empirical proposition that a is an F: (1) an atomic one for retrieving a and the property F, respectively; (2) a complex one for obtaining a set from a property and applying the set to an argument; (3) another atomic one for retrieving a world and a time, respectively, as empirical indices, and another complex one for obtaining a truth-condition.

The overall picture that emerges is that a structured proposition is procedurally structured out of sub-procedures, some of which are atomic (of one-step) and some of which are complex (of multiple steps). The theory I am deploying is at heart an elaboration, by way of procedures and logical types, of a function/argument logic in the vein of Frege-Church-Montague and as encapsulated in the *Tractatus* (§5.47), that whenever there is compositeness, argument and function are present.

This is intended as a discursive paper in foundational philosophy of language. The formal foundations and technical details can be found in my (ms.), and in Duží et al. (2010) and Tichý (1988). Much of my thinking about structure is indebted to Tichý (1995).

The rest of this paper is organized as follows. Section 1 provides conceptual and historical preliminaries concerning predication and unity. Section 2 outlines a subjectivist or act-theoretic or pragmatic conception of predication and how predication serves as the unifier of atomic propositions. The theory in question is Hanks's theory of propositions as types of action. Section 3 outlines an objectivist or operation-theoretic or semantic conception of predication and how predication serves as the unifier of atomic propositions. The theory in question is Tichý's theory of propositions as complex objective procedures, which is called Transparent Intensional Logic (TIL). Section 4 briefly compares these two theories.

1. 'Such Wholes Are Always Propositions'.

Post-WWII semantics has known three prevalent theories of propositions.[1] The first was propositions as functions from possible worlds to truth-values, or equivalently, satisfaction classes of possible worlds. The second was propositions as sequences, i.e. enumerations of entities and perhaps also operations. The third was, and still is, propositions as structures. The first two construe propositions as set-theoretic or model-theoretic artefacts. The third one falls outside this paradigm and calls instead for mereology.[2] We are currently going through a shift from sets and their elements to wholes and their parts. This reorientation throws us right back to problems that Russell was struggling with around the turn of the century and which Frege had struggled with before him. The problems, in both cases, bear on the logical interaction between properties, relations and objects. I maintain (without argument for now) that any account of this interaction faces two constraints. One is that interaction must not degenerate to a mere list of entities and operations. The other is that interaction must not unleash an infinite regress.[3]

Let us get more specific. How do an object and a property combine so as to form the proposition that the object has the property? Why, that is easy. Just add *Zipping*. Merricks (2015: 154) explains that there is nothing to explain:

> Nothing explains why *Zipping* works this way. That is just how *Zipping* works. And neither *Zipping*'s existence nor its holding between [object] O and [property] F depends on anything that agents do. This reply concludes that the proposition *that O is F* is the state of affairs of O's being related by *Zipping* to F.

Merricks (2015: 154, fn. 28) continues:

> Mark Johnson seems to endorse a *Zipping*-based account of the unity of the proposition, although he calls the relation in question 'predication', rather than (the admittedly unflattering) '*Zipping*'.

The passage in Johnson (2006: 684) Merricks quotes from states, inter alia, that:

---

1   I am glossing over the constructivist conception of propositions as constructive sets of constructive proofs (so-called constructions). Martin-Löf's type theory is a both conceptually and formally elaborate example of constructivist propositions. I am also glossing over the conception of propositions as primitives. Bealer's intensional algebra is an elaborate example of this tack. Finally, I am glossing over propositions as functions from impossible worlds to truth-values, as found in Hintikka, Priest, etc.
2   Cf. Gilmore (2014, esp. 182ff).
3   See my (2012), (2012a) on the *list* and *regress* objections to alleged structures.

(…) we can identify the proposition that *a* is *F* as the predication of *F*-ness of *a*. We may think of that as a complex item built up from *F*-ness and *a*, by way of the relation of being predicated of. So what it is for the proposition that Aristotle liked dogs to exist is for liking dogs to be *predicated* of Aristotle.

Merricks in effect challenges Johnson to provide an account of predication that will make predication less magical and mysterious than afforded by *Zipping*. How will adding predication to *a* and *F* yield the proposition that *a* is an *F*? Johnson himself brings up two different ways of going about this, without providing details so as to present a theory rather than a project. One way is the *subjectivist* approach pivoted on subjective acts of predication. The other way is the *objectivist* approach pivoted on an objective link between objects and properties. The bifurcation between subjectivism and objectivism is rooted in his disambiguation of the act/object ambiguity he ascribes to the term 'predication' (Merricks 2006: 684).[4] Here is how Johnson brings the two approaches together, stating a preference for objectivism:

> (…) when I perform the act of predicating *F*-ness of some individual *a*, I thereby relate myself in judgment to an objective entity, the predication of *F*-ness of *a*. I judge true the predication of *F*-ness of of *a*. But this predication of *F*-ness of *a* is just the proposition that *a* has *F*-ness.

Here Johnson equates proposition and predication. The proposition that *a* is an *F* is the predication of *F* of *a*. The salient point is the status of predication. Once we understand predication, we shall understand what a proposition is, including what the unity of at least atomic propositions amounts to. The objectivist needs to

---

4 Objectivist theories of predication encompass both so-called 'Fregean' and 'Russellian' propositions. Subjectivist theories would encompass Russell's multiple-relation theory and the act-theoretic theories of Soames and Hanks. In fact, given the free-wheeling manner in which the monikers 'Fregean' and 'Russellian' are bandied about as labels for two clusters of theories of propositions, I presume we are entitled to introduce the moniker 'Husserlian' for subjectivist or act-theoretic theories of predication and propositions (bar the naturalism). Also it might be interesting, in a later study, to compare act-theoretic against constructivist propositions. 'Fregean', as opposed to 'Russellian', seems to signal that the basic notion is *function* rather than *relation* and that the parts are exclusively *senses* rather than *objects* or a mixture of individual objects and universals (though everything in Russell's universe, at the highest level of abstraction, is a *term* in his non-linguistic sense thereof). It is helpful to use these labels as a quick guide to the territory, but beyond that they soon become strained. There are no ready-made, fully implementable Fregean or Russellian theories of propositions.

make good sense of the 'objective entity' that is the predication of F of a.[5] The subjectivist must explain how a proposition can be equated with an act of predication.

As an avowed advocate of hylomorphism (perhaps using this term lightly) for propositions, Johnson seeks a principle of unity that will unify a and F, noting that their mere juxtaposition is insufficient. Johnson (2006: 659) emphasizes that the principle is not a further part of what it unifies. It is obvious why not: if a unifier is an entity alongside the entities it is designated to unify then we are triggering a regress. Johnson thinks of predication as the sought-for unifier, but in turn thinks of predication as a relation (cf. 2006: 685). I recommend we not think of predication as a relation. Russell went down this path, which was paved with Bradley's regress.[6] Instead I recommend thinking of predication along functional lines. When the details are added, predication emerges as an instance of the objective logical procedure of functional application.

I grant that one may look upon the problem of unity in the same spirit as Wordsworth gazed at the rainbow in awe and joy, with natural piety. Thus is García-Carpintero's Romantic recommendation (2010: 288). He in effect recommends taking unity as a primitive in one's account of propositions: there is going to be a niche for the notion in the edifice, but the notion is not going to be analyzed. This is a theoretical option, of course, but my view is rather that analyzing unity

---

5 I restrict myself to those *atomic* propositions that are monadic predications without adjacent secondary predication. This rules in the proposition that Mary is happy and rules out the proposition that Mary leaves happy. These minimal atomic propositions demand an account of how individuals/objects and properties (monadic intensions) interact so as to form the proposition (medadic intension) that something or somebody has a property. I neglect polyadic intensions (relations of arity two or more) as predicables, since the basic question remains the same, namely how two disparate objects form a third kind of entity. I also neglect properties of empirical properties, as expressed by "The megalodon is extinct" or "Kindness is common". My answer is the same in all three cases: by means of predication. *Complex* propositions are a different kettle of fish. There we already have propositions as input. The question is instead how to form a new proposition from one or more existing propositions. There are type-theoretic complications, for instance, that the truth-functional connectives operate on truth-values (propositions-in-extension) rather than propositions or that entailment, which does operate on propositions, will output a truth-value rather than a proposition. But the basic idea should be clear, that there is a host of propositional operators to choose from. See also Merricks (2015:197ff) concerning atomic and complex propositions. My (2012a) offers an embryonic account of the unity of complex propositions.
6 Johnson (2006) is short on the details, though, so I cannot say for sure whether he is off to a regress.

is a fruitful way to address topics like predication and structure. I find that the minute one starts talking about *structured* propositions then one must explain how one's structures unify many things into one thing such that what is described is a *system*. There is one or more mechanisms, or principles of unity, to be unearthed. Conversely, if the notion of unity is not going to be analyzed then either one goes in for structured propositions whose structure is taken as a primitive or for propositions as primitives or for propositions as set-theoretic entities.

The following three tenets summarize my position.

- *Predication is a logical operation.* At the same time human acts of predication are a fact of linguistic life. Only the operation is conceptually prior to the act. This marks a preference for semantic realism over semantic anti-realism.
- *Predication is an instance of the procedure of functional application.*
- *Predication is the unifier of atomic propositions.*

I can imagine three conceptions of predication. They tally neatly with the three parts of semiotics, except that 'logical' has replaced 'semantic', though I am happy to say 'logico-semantic' or simply 'semantic'.

- *Syntactic conception.* The cornerstones include notions like copula, sentence structure, declension, NPs and APs.
- *Pragmatic conception.* The fundamental idea is that a speaker or thinker predicates a property of an individual when executing a speech act or an intentional act.
- *Logical (semantic) conception.* The leading idea is that a logical operation applies to one or more operands. In the present case the input operands are a property and an individual. From there one can go in various directions. One may identify the output entity with the proposition that the individual has the property, or one may identify the proposition with the very operation that takes the property and the individual to an output entity, which again may be either a truth-condition or a truth-value.

Williams (1980) exemplifies the syntactic conception. His concerns are all about filtering well-formed from ill-formed predicative syntactic structures (in English, as it happens). Predication, syntactically speaking, is obviously a *relation* between noun and predicate phrases, where predicates can be drawn from a multitude of grammatical categories. I won't be considering the syntactic conception any further here. Hanks (2015) and King et al. (2014) exemplify the pragmatic conception. Frege and TIL, e.g. Jespersen (2008), exemplify the logical conception, like we just saw Johnson (2006) do.[7] So does the Russell of (1903), but not the Russell of (1910).[8] Lenci (1998:251) points out that Frege drove in a wedge between syntactic and semantic (logical) predication. The difference is between whether a sentence or a proposition is formed by means of predication: "semantic composition is not terministic [i.e. term-based, as in Aristotle] but functional" (Lenci 1998: 249). The two part company because the copula, for Frege, is semantically and logically idle. The historical rift is summarized by Lenci (1998: 237) thus:

> While for Mill [who adhered to the Aristotelian and medieval subject/copula/predicate structure] predicates need the copula to be predicated of a subject, for Frege predicates express concepts [properties], and concepts are intrinsically predicative, i.e. functional [i.e. unsaturated, with gaps welcoming arguments for saturation].

Semantic composition is a matter exclusively of the operation of functional application in Frege, just as in Generative Grammar.

How does predication bear on propositional unity? As I understand the unity problem, it has two aspects, one metaphysical and the other semantic. I agree entirely with Johnson's (2006: 683) characterization:

> The problem of the unity of the proposition is then two-fold: How is it that items like Aristotle [a *topic*] and the property of liking dogs [a *predicable*] make up some third complex item, the proposition that Aristotle liked dogs, and how is it that a complex item like this, unlike the sum or the set or the group or the sequence of the constituents, is a nonarbitrary bearer of truth value?

The first aspect is the *mereological*, hence *metaphysical* problem of many-into-one composition: how do disparate entities like an individual and a property hook up so as to form a third kind of entity, a proposition, rather than just an enumeration

---

7   King (2007) and later moves from syntax through pragmatics to semantics. García-Carpintero (2010: 287–288) levels what amounts to a *Zipping* objection against King's account of unity, an account which seeks to extend syntactic predication to semantic predication.
8   Levine (2002) is a highly recommendable study of Russell and Frege on the structure of predication and propositions. Levine makes a compelling case for the interpretation that a *Gedanke* is not the sort of fine-grained, structured entity it is commonly taken to be. Rather it is coarse-grained, being individuated only up to logical equivalence, and the finesse to do with fine-grained individuation needed for cognitive significance resides instead in the different ways of arriving at a *Gedanke*. Briefly, a *Gedanke* is a functional value rather than a function in the pre-mapping sense of 'function'. See Duží et al. (2010, p. 3). See also my (2015).

of entities? This is the problem of *interaction* among the constituents.⁹ It is conceivable that a decomposition of the proposition that *a* is an *F* will contain only *a*, *F* (or presentations thereof) as stand-alone constituent parts, but this is not to say that there is not going to be more than that to that proposition. The parts need to 'talk to' one another. This point is lifted straight from Russell (see below).

The second aspect is the *semantic* problem of how a proposition gets to be or have or yield the *truth-condition* it does. A kindred way of framing the problem is in terms of how a proposition may be a vehicle of *representation*, e.g. the representation of *a* being an *F*, or *F* holding of *a* (depending on whether we wish to highlight the property or the individual). It is crucial that non-truths can be represented, and can be represented on an equal footing with truths. Therefore predication must be non-factive. I suggest contrasting predication with instantiation, such that if *a* instantiates *F* then the predication that *a* is an *F* is true, whereas if *a* fails to instantiate *F* then this predication fails to be true. Briefly, true predication coincides with instantiation. A different way of raising the representation problem is to raise the question of how propositions can be *truth-apt*, i.e. suitable as bearers of truth-values.¹⁰ It should be fairly clear how the metaphysical and the semantic aspect hang together. The representation of *a* as being an *F*, or *F* holding of *a*, presupposes interaction between *a*, *F*, and an account of this interaction is an

---

9   Set theory, at its 'naïve' inception, faced a similar many-into-one problem: how can multiple objects be a unit, such that these elements are considered together (cf. the Italian and French words for *set*, 'insieme' and 'ensemble') and such that the set may have properties its elements lack, like being fifteen-membered. However, as the elements do not interact within the set, set theory faces only the 'outward-looking' problem of fixing the perimeter of a set in order to seal off its *elements* from everything else in the universe. The sort of mereology suitable for structured propositions faces the additional 'inward-looking' problem of getting the *parts* to 'talk to' each other so as to gel into a whole.

10 See Liebesman (2015). It is obvious that the metaphysical unity problem arises only for structured propositions. It does not arise when propositions are either black-boxed primitives or set-theoretic entities, such as sets of possible worlds (a satisfaction class) or functions from possible worlds to truth-values (a characteristic function). But the semantic unity problem arises for any sort of propositions, in that any theory of propositions must explain how its propositions relate to representation and truth-aptness. Only I would hesitate to describe this question as a question of *unity*, if divorced from structure. Solely theories of *structured* propositions must explain how their propositions are unified and structured so as to be truth-apt and representational. Theories of unstructured propositions simply skip this step.

account of the predication of *F* of *a*, which in turn is an account of the proposition that *a* is an *F*. In this paper I focus on the mereological aspect of unity.

Russell was clear that wholes are one thing and sets, or collections or classes, another.[11] Russell (1903: 140–141) distinguishes between two sorts of wholes, *aggregates* and *unities*:

> An *aggregate*] is completely specified when all its simple constituents are specified; its parts have no direct connection *inter se*, but only the indirect connection involved in being parts of one and the same whole.

There *is* a difference between such a whole and all its parts, even if the difference is minimal. I note in passing that the identification and individuation of an aggregate obeys the principle (A) ('Analysis') of Levine (2002).

> But other wholes occur, which contain relations or what may be called predicates, not occurring simply as terms in a collection, but as relating or qualifying. *Such wholes are always propositions*. These are not completely specified when their parts are all known. (*Italics added*.)

Russell subsequently presents his famous (notorious?) example of the proposition that *A* differs from *B*, and his enigmatic conclusion is:

> The fact seems to be that a relation is one thing when it relates, and another when it is merely enumerated as a term in a collection.

I understand this difference to be in the same ballpark as the difference between *displayed* and *executed* procedures that I will broach below.

The main difference between aggregates and unities is this:

> Each class of wholes consists of terms not simply equivalent to all their wholes [which marks the difference between wholes and sets]; but in the case of unities, the whole is not even specified by its parts. For example, the parts *A*, greater than, *B*, may compose simply an aggregate, or either of the propositions [denoted by] "*A* is greater than *B*", "*B* is greater than *A*". Unities thus involve problems from which aggregates are free.

---

11 Both Merricks (2015: Chs. 3–4) and Johnson (2006) offer critiques of various set-theoretic accounts of complexity and unity. See also my (2013) and (2003) for objections to even modeling propositions as tuples, however many operations are added as designated unifiers. Let me say a few words about one objection. The fact that tuple theories must deem the choice between $\langle a, F \rangle$ and $\langle F, a \rangle$ arbitrary goes to show that they have nothing interesting to say about the interaction between *a* and *F*. For this reason a tuple does not qualify as a structure or a system. A tuple is just a line-up, and as such suitable as a functional argument or value. Still, the *parts* of a structured proposition can be lined up as the *elements* of a tuple that mirrors the sequence in which they already occur in the proposition.

So we have been duly warned. Whereas aggregates are a half-way house between sets (including sequences) and unities, unities are characterized by a *connection inter se* among their parts – and accounting for connection inter se within such wholes as are propositions is not unproblematic.¹² For comparison, Merricks (2015: Ch. 4) adumbrates (and dismisses) a theory he calls the 'merely mereological' theory of unity. It is, in effect, extensional in the sense that a whole having such-and-such parts is identical to any whole with the same parts. This theory maintains that wholes "are united in exactly the same way – namely, only by composing something" when their parts are the same. No further distinction is in stock, e.g. in terms of interaction: composition is black-boxed. He continues (Merricks 2015: 146), "the only structure [the defenders of the merely mereological] countenance for propositions is mereological". I suggest that the lesson is that we must broaden the mereological idiom, from the extensional one countenanced by Merricks to a hyperintensional one, which will enable hyperintensional distinctions among propositions that have the same parts, and do so in a non-primitive, non-black-boxed manner, namely in terms of logical procedures.

I will outline a theory of *connection inter se*. Connection is never a relation, whether busy relating its relata or just on display as one item alongside other items. Connection inter se is a logical procedure, and the procedure in question is the one of applying a function to an argument. The procedure is not executed: we do not progress to the value of the function at the argument. The procedure is displayed as a procedure in its own right. Therefore properties can be predicated of the procedure rather than of the entity it produces. The adjacent type theory tells us what type of object would be returned as value were we to apply a function of a particular type to an argument (or a string of arguments) of a particular type. The distinction between procedure and product (the result of application) is crucial.

---

12   Gaskin (2008: 345) says, "what stops a proposition from being a *mere* aggregate of entities […] is that the proposition unfolds into an *infinite* aggregate […]", and goes on to argue that Bradley's regress is not vicious, but explanatory and manageable, and 'the metaphysical ground of the unity of the proposition'; e.g. "The regress unpacks the original proposition into an infinity of further propositions …], but it does not present us with a series of discrete epistemic tasks […]. […] its unity depends on the presence […] of the members of the regress: […] if the regress did not get going, or if it faltered at some point […,] the proposition in question would not be unified, but would fall apart into a mere aggregate" (2008: 352). The impression I come away with is that Gaskin's unity is an aggregate *de luxe*. García-Carpintero (2010: 288) regrets that Gaskin appeals to a regress that he claims not to be vicious in a way that he (García-Carpintero) is unable to make consistent sense of.

The procedure is the bearer of the properties of being complex and structured, whereas its product is a set-theoretic entity (a truth-condition).

The challenge that Merricks (2015: 143, fn. 20) puts to the Fregean is that their senses should not be on a par with *Zipping*. That is, when Fregean senses are invoked it must be shown *how* they interact. I identify Fregean (well, 'Fregean') senses with procedures, and I shall describe, in prose, how they interact. The resulting theory is that a structured hyperproposition is a system of interlocking (sub-) procedures.

2. Acts of Predication

Predication and assertion walk hand in hand, according to Hanks (2015). Hanks's theory of propositions and predication is at heart a pragmatic one, for propositional *content* is drenched in assertoric *force*.[13] The proposition that *a* is an *F* is identified with a complex act type consisting of three distinct types of sub-acts (cf. Hanks 2015: 77). The action of uttering the sentence "*a* is an *F*" decomposes into the sub-act of referring to *a* by means of '*a*' and the sub-act of expressing the property *being an F* by means of '*F*', while the overarching act is to perform the predication of *F* of *a*. Hanks says (2015: 163):

> Predication [in contrast to judgment] can be viewed as a multiple relation. We can take predication to be a poly-adic relation that a subject bears to various, disconnected entities.

Hanks's theory is not stymied by what undermined Russell's multiple-relation theory.[14] A proposition, whether true or not, is a type of action that exists beyond being tokened by an agent in an act of predication. Also Hanks can easily accommodate untrue propositions, for it is an option for an agent to predicate *F* of *a* while *a* fails to instantiate *F*. Russell was never able (or perhaps willing) to accommodate false propositions other than at most as ideal or subjective entities.

But now that the centre of gravity has shifted to subjective acts of predication, Hanks faces a problem not unlike Russell's with propositions anchored in subjective acts of judgement. The default is that an act of predication comes with assertoric force: to predicate *F* of *a* is equipollent to asserting or holding true that *a* is an *F*.

---

13 The dividing line in the act-theoretic community is between whether acts of predication are committal or not. Hanks's default position is that they are; Soames's that they are not. When a speaker or thinker is predicating *F* of *a*, are they ipso facto holding it to be true that *a* is an *F*? Reiland (2012) offers a discussion of this divide, finding trouble with both. Hanks (2015: 35ff) argues that Soames's non-committal position is incoherent.

14 Hanks (2015: 161–166) offers a comparison of his own position with Russell's multiple-relation theory.

Therefore predications without assertion become conceptually problematic, and so Hanks's architecture needs to make room for them. He is alert to the problem, of course, and I venture to say that solving it is the greatest challenge facing his position. The tenet that predication is assertion is a bold move forward, but predication without assertion is required as well. His theory needs a mechanism for switching assertoric force on and off while leaving the very acts of predication unscathed. His solution is called *cancelled predication* (Hanks 2015: 99):

> (…) cancelled predication is not predication minus commitment. There is no such thing as partial, non-committal predication. Cancelled predication is more than predication, not less.

On the face of it, this claim looks peculiar. The modified predicate 'is cancelled predication' seems to exemplify privative modification. For instance, a cancelled flight is a non-flight, so cancelled predication would seem to be non-predication and most certainly not more than predication, but much less than or even incommensurate with predication. Be that as it may, Hanks's default case of predication-*cum*-assertoric force is actually the rare case, for it includes only atomic and conjunctive propositions.[15] In all other cases the predicative acts are set within so-called cancellation contexts, in which assertion has been divorced from predication. Despite many suggestive remarks about cancelled predication, we still need to be told what the underlying mechanism is for switching assertoric force on and off. It is not sufficient to state *that* an act of predication has assertoric force

---

15 I was puzzled to learn that factive attitude contexts are also cancellation contexts. Hanks (2015: 151–152) says, "the verb 'believes' is responsible for creating a cancellation context for the embedded act of predication. (…) [Are] there attitude verbs that do not generate cancellation contexts [?] The verb 'knows' seems like a good candidate. Perhaps 'knows' does not create a cancellation context. If not, then someone who asserts that Obama knows that Clinton is eloquent would also assert that Clinton is eloquent. (…) Unfortunately this won't work." Hanks's reason for having 'knows' create a cancellation context goes via an analysis of "Obama knows *whether* Clinton is eloquent". The same analysis (whatever the details) cannot, of course, carry over to "Obama knows *that* Clinton is eloquent". *Knowing whether* is not factive, *knowing that* is, and "[i]f the use of 'knows' cancels in this case, then uniformity demands that it cancels in all cases." This seems like a *non sequitur* from uniformity. For uniformity, *knowing whether* and *believing that* should be grouped together as non-factives, and *knowing that* together with the other factives. This seems in the spirit of Hanks's theory: when Obama knows that Clinton is eloquent then Obama thereby assertorically predicates eloquence of Clinton. See Duží et al. (2010: 453–458) for the logic of *knowing whether*, including Def. 5.2. and the six-way disambiguation of "*a* knows whether Scott is the author of *Waverley*".

outside of a cancellation context and lacks assertoric force inside a cancellation context. This is important not just to account for, e.g., disjunctive propositions, but also for the traffic in and out of cancellation contexts: see, for instance, the cursory remark at (Hanks 2015: 111) on predicating F of a for a *reductio*. Put polemically, the claim cannot merely be that predication is assertion, except when it is not.[16]

If a proposition is a predicative act and if assertoric force is interwoven with propositional content, will cancellation detract from the content of a proposition? One hopes not, for cancellation is a vehicle of contextualization and the context within which a proposition occurs should not impinge on its content.

The official doctrine is that cancelled predication is predication without assertion. But in my (2012) I voiced the worry that cancellation might perhaps undermine the unity of cancelled predications, i.e. the unity of most kinds of propositions. The doctrine is that a proposition within a cancellation context is no longer used to make the assertion that a is an F. Instead the proposition occurs as an argument that is hospitable to propositional operations. So how are *a*, *F* – or the sub-acts of singling out *a*, *F* – kept together? An agent is required to keep them together by means of an act of predication. Only it is an act of predication without assertoric bite. How does that work? When propositions do have assertoric force, it at least seems intuitive enough that the speech act of coming forward with the claim that *a* is an *F* goes some way toward explaining how *a*, *F* hook up. One points out that somebody has performed the act of asserting *F* of *a*. This explanation does not apply to cancelled predications (propositions occurring within the toxic environment of cancellation contexts). So what is the mechanism, if not the act of

---

16 Moreover it is not a foregone conclusion that the predicative act *that a is an F and b is a G* is tantamount to two separate acts of predication: see Hanks (2015: 105). To assume that predication distributes over conjunction is parallel to assuming, e.g., that knowledge distributes over conjunction. Comparing act-theoretic predication to knowledge is reasonable because also the former induces an attitude context, so issues like granularity and closure crop up. It is contentious, however, whether, as a matter of logical or analytic necessity, someone who knows *that A and B* must ipso facto know *that A* and know *that B* individually. It is conceivable that an agent masters conjunction introduction (or simply starts out with a conjunctive proposition) but not conjunction elimination and so is unable to detach individual conjuncts. Or even if the agent masters conjunction elimination, it is still not obvious that the attributer is allowed to attribute individual conjuncts to the agent. What is at stake is overgeneration of knowledge; similarly with overgeneration of predicative acts. What Hanks would need to demonstrate (rather than just assume) is why it should be a matter of course that predication distributes over conjunction.

assertion, that keeps a, F (or their sub-acts) together? I am not sure Hanks (2015) contains a convincing answer to that.

Nor am I sure Hanks's notion of cancellation (cancelled predication, cancellation context) is compatible with his equally important tenet that force is integral to content. The effect of cancellation is to blot out force while preserving content, thus restoring the force/content distinction Hanks is adamant to tear down. The deep problem is not so much that Hanks's position on cancelled force appears to be verbally indistinguishable from Soames's, but that it is not even clear what Hanks's content devoid of force would be (whereas it is reasonably clear what Soames's content without force is).

In fact, the question of unity crops up whether or not we can invoke the category of assertoric speech acts. For also when somebody performs the act of asserting F of a is it legitimate to ask how that works. In a nutshell, what is the logic (the logical operation) of assertion? Whichever it may be, it is not built into Hanks's pragmatically informed semantics. The view that acts of predication, whether permeated with assertoric force or not, undergird propositional unity *presupposes* an answer to the question what the underlying logical mechanism is.

3. Predication as a Procedure

Here is a quick survey of my position. Predication – as opposed to instantiation, set membership, etc. – is the operation whose logic is the instruction to apply a function when the extension of a property and an individual are juxtaposed so as to obtain a truth-value (not exclusively True) from it. This operation must be integral to the structured proposition. The opposite would be to project the operation onto the proposition from the outside.

What unifies the atomic proposition that a is an F is not, and must not be, an ingredient alongside the ingredients a, F (or their senses) on pain of regress. What does the unifying is a structure. I identify structures with procedures. When empirical, the proposition that a is an F comes with two unifiers. One is the inner unifier that takes care of predication. The other is the outer unifier that instructs us on how to ascend from a truth-value to a truth-condition.

Let me contrast my view with King's. King (2007: 34) talks of syntactic concatenation as giving the instruction to map an object o and a property P to true at a world iff o instantiates P at that world:

> This instruction has two crucial features. First, it involves a specific function f: the function that maps an object and a property to true (at a world) iff the object *instantiates* the property (at the world). Call this function f *the instantiation function*. Second, the instruction instructs that f *is to be applied* to the semantic values of the expressions at the

left and right terminal nodes [o and P, resp.] (and a world) to determine the truth value of the sentence (at the world).

I differ on three accounts. First, I do not speak of an instantiation function, but a predication function. King's f and my predication function are both defined (typed) to return truth-values as functional values, but instantiation is just one half of predication. (This is mainly hair-splitting, though, in the light of the second 'crucial feature'.) Second, King's f appears to be defined to take a world to a function from a pair consisting of an individual and a property to a truth-value. My predication function presupposes that the property has already been extensionalized to yield a set. If King thinks of properties as functions from worlds to world-relative satisfaction classes, I am not sure I understand the (type-theoretic) details of how individuals and properties and truth-values are to hook up. Third, and most importantly, I assign no instructions to syntactic concatenation or try to visit a syntactically rooted instruction upon a logico-semantic entity such as a structured proposition. Instructions (procedures) do not emanate from syntactic entities or pragmatic acts, but are introduced straightaway, as entities in their own right. My methodology is top-down, from abstract entities, rather than bottom-up, from acts and actual syntax. This difference in methodology is probably due to the fact that I do not share the naturalistic aspirations that inform the act-theoretic conception of predication and propositions.

Let me also contrast my view with Soames's. Soames (2010: 114) draws this distinction:

> (…) to apply [function] f to [argument] a is not to calculate the value of f at a, but merely to take a as an argument of f.

In the idiom of procedural semantics, one can entertain the procedure of applying f to a without executing the application. The ability to distinguish between that procedure and its result, $f(a)$, is fundamental for the fine-grained individuation of Soames's propositions. For instance, despite the equivalence ($6^3 > 14^2$) ≡ (216 > 196), one thing is to predicate *being larger than* of the pair $\langle 216, 196 \rangle$, another thing is to predicate *being larger than* of $\langle 6^3, 14^2 \rangle$. Soames concludes that the position $\langle \alpha \ldots \beta \rangle$ in the context $\langle X$ predicates P of $\alpha \ldots \beta \rangle$ is non-extensional, in that equivalents cannot be validly substituted.[17] The 'cognitive act' of applying a function is 'akin to the act of predicating a property of something' (Soames 2010: 114). Hence predicative event types come out fine-grained. A predication context is fine-grained, because it is in effect an attitude context. My predication contexts,

---

17 Cf. Soames (2010: 112).

by contrast, are extensional, because the nucleus of predication is the instruction to apply a characteristic function to an argument. Nonetheless, there is more to predication than this simple instruction to apply a function in the case of empirical propositions. For one thing, the set has been obtained by extensionalizing an intensional entity (a property) and the truth-value resulting from the application is abstracted over to obtain another intensional entity, namely a truth-condition. Furthermore the instruction to obtain a truth-value and the instruction to obtain a truth-condition from it are embedded within a structured proposition that is a complex procedure. What we are looking at here is logical traffic between the extensional, the intensional, and the hyperintensional level.

Before proceeding, I should allay the following worry. When two different properties F, G share the same extension at the world/time of evaluation, how are we to distinguish between the predication of F and of G? We are able to distinguish because the same set $F_{wt} = G_{wt}$ has been obtained via two different routes, either starting from F or from G. The respective procedures of functional application will exhibit this difference. There is 'no backward-road' from the resulting characteristic function to those two procedures. The answer I just gave cannot be given if instead some particular set $\{a, b, c\}$ is offered up as functional argument. There is one more reason why $\{a, b, c\}$ is inappropriate when it comes to predication. If predication reduces to set membership then the predication of *contingent* properties is rendered impossible for the simple reason that it is necessarily true that $a \in \{a, b, c\}$ and necessarily false that $a \notin \{a, b, c\}$ or that $a \in \{b, c\}$.[18]

Tichý (1988: 63–65) defines five different sorts of procedures, having already introduced variables as a sixth sort of procedure. He calls them *constructions*.[19] The definition of the sort of construction he calls *Composition* is a definition of the procedure of functional application. Where M is a construction of a mapping and A a construction of an argument of the constructed mapping, $[MA]$ is a Composition and not a functional value. $[MA]$ decomposes into these three subprocedures or steps:

- Execute construction M to obtain a function m typed to take such-and-such a type of entities to such-and-such a type of entities
- Execute A to obtain an entity a of such-and-such a type

---

18 For reference, this argument is valid: (1) [$^0F_{wt}\,^0a$]; (2) [$^0 = {}^0F_{wt}\,^0G_{wt}$]; ∴ (3) [$^0G_{wt}\,^0a$]. Still, if $F \neq' G$ then (1) and (3) qualify as predicating two different properties.

19 Programmatically put, Tichý's variables are Russellian rather than Tarskian as far as their metaphysical standing goes – objectual rather than linguistic. Logically speaking, they are like Tarski's variables.

- Apply function *m* to *a* as argument to obtain an entity of such-and-such a type.

It is arbitrary in which order one executes *M*, *A*, when it is fixed that *m* is to be applied to *a*. [*AM*] is also a Composition, although a type-theoretic monstrosity, because it is the procedure of applying an argument to a function. Some executions run *in parallel*, *in casū* the respective executions of *M*, *A*. Other executions are *serial*, especially when one construction must be executed to yield a functional value which subsequently becomes the argument of a new function. My initial flow-chart metaphor was intended to capture the serial character of executions and the value-to-argument transformations.

It bears repetition that Composition is the very *procedure* of functional application. Had I simply said 'functional application' then that would have been ambiguous between the *result* of the application (a functional value) and the *procedure* which, if executed, has that result.[20] When I talk about Composition, it may be helpful to imagine a function and an argument both being lined up, primed for application, but without the execution of the procedure taking place. The step from procedure to product is not made. This distinction is important because what is philosophically relevant is the procedure and not its result. The resulting functional value is namely a truth-value, of next to no philosophical import, for it reflects whether an extra-semantic, empirical truth or an empirical falsehood obtains. When I say that functional application is what undergirds predication and unity, it is obviously Composition – or in general, extra-TIL terms, the procedure of functional application – that is intended.

So Composition is the logic of predication. But what makes some, though not all, instances of Composition instances of *predication* rather than of something else?[21] My simplest argument is: what could it be if not predication? It cannot be instantiation, if instantiation is understood to be factive. The resulting value is a truth-value and not the truth-value True. My philosophical view is that the logic of predication is exhausted by the juxtaposition of a suitably typed function and argument accompanied by the instruction to apply the function to the argument to obtain a truth-value. Which truth-value is obtained depends on whether the argument entity instantiates or lacks the property that has been extensionalized. Saying this much commits me to the view that any such juxtaposition must count

---

20 Tichý allows the composition of a function with an argument that fails to return a value. I am suppressing this complication here. See Duží et al. (2010: 276–278).
21 Rafael De Clercq raised this question after my talk at Lingnan in September 2015. I had anticipated the question in my (2008). The answer offered above is expanded on in my (ms.).

as a case of predication. But I readily admit that Composition – even when the function and the argument and the value are as just described – is not inherently predication. An obvious alternative would be to introduce an operation specially earmarked for predication. But I am not in favour of that, because it multiplies single-purpose operations.[22]

The dual of Composition is *Closure*, which is the procedure of functional abstraction. They are duals because Closure 'builds up' (or 'declares') functions while Composition 'breaks down' functions by 'applying' them. The two auxiliary 'feeder' procedures are Trivializations and variables. They are atomic procedures which retrieve a specific object and an arbitrary object, respectively, from a domain of objects of a particular type.[23] All four procedures are *ideal*. They may or may not be executed, while other ones cannot possibly be executed, so the procedures count as Platonic entities.

To be specific, consider the empirical proposition that a is an F. A proposition qualifies as empirical as soon as it is true at only some world/time pairs that a instantiates F. Empirical properties can be predicated of individuals only relative to worlds and times, rather than absolutely. For instance, let us consider the proposition that Miles smiles. The nucleus of this proposition is the predication of *smiling* of Miles. We start out with an instance of the Composition [MA] together with applications to a world and a time.

- Execute the Trivialization of *smiling* to obtain the property of smiling
- Execute the variable w to obtain a world
- Apply *smiling* to the world of evaluation to obtain the characteristic function of the chronology of those who are smiling at that world, where a chronology is here a function from times to sets of individuals
- Execute the variable t to obtain a time
- Apply the above chronology to the time of evaluation to obtain the characteristic function of the set of individuals who have the property of smiling at that time

---

22 My (2008) argues against Bealer's use of single-purpose predication operations (of which he has two).
23 It does not matter, for the points I want to make here, whether 'a' is a proper name or a definite description or a pronoun (whether anaphoric or free), as long as an individual is provided. Nor does it matter whether F is a basic property or a property obtained by modification, property conjunction, etc., as long as a property is provided. The important thing is the interaction between a and F_{wt}. However, ^0a is the Trivialization of a and as such the meaning of any proper name denoting a. My (ms.) explains Trivialization in detail.

- Execute the Trivialization of Miles to obtain the individual Miles
- Apply the characteristic function of the above set to Miles to obtain a truth-value, according as Miles smiles or not at the world and time of evaluation.

At this point we have obtained a truth-value. The application of the characteristic function of the property *smiling* to Miles returns either True or False, according as Miles is an element of the set of smiling individuals at the world and time chosen for evaluation.[24] We have reached extensional rock bottom. From here we begin to ascend to an intensional entity. Exit Trivializations and Compositions. Enter Closures and variables.

- Abstract over the resulting truth-value by means of a variable ranging over times to obtain a function from times to truth-values (i.e. a chronology of truth-values)
- Abstract over the above chronology by means of a variable ranging over possible worlds to obtain a function from worlds to such chronologies.

At this point we have obtained a truth-condition: the set of world/time pairs at which it is true that Miles smiles.[25]

The above bullets taken together constitute a decomposition of a compound procedure into its sub-procedures. When we go in the opposite direction, by recomposing the complex procedure, we obtain a nine-step structured hyper-proposition. It is a sequence of steps, but it is more than that. Above all, a procedure, unlike a sequence, can be *executed*. The feeder procedures (Trivializations and variables) describe how to obtain entities, and the operational procedures (Compositions and Closures) describe what to do with the entities thus obtained.

The above Closure, importantly, occurs in *displayed* mode. We are looking at the procedure (a hyperproposition) and not at its product (a truth-condition). The procedure is 'suspended' or 'stretched out' and not 'collapsed'. In virtue of the fact that this Closure occurs displayed, it can itself occur as the argument of a function. Therefore properties can be predicated of it. I already predicated the

---

24 Let me clarify one point. I am not saying that we obtain the entire population of smiling people at w, t, in order to check whether Miles is an element of this population. Rather, for *any* individual, we empirically try to decide whether a given individual qualifies, at w, t, as a smiling individual. The difference between *all* and *any* is important here, and this difference is secured by using the λ-abstractor.
25 For reference, the construction is: $\lambda w \lambda t$ [$^0Smiles_{wt}$ ^0Miles]. Or if we spell out all the instances of Composition: [λw [λt [[[^0Smiles w] t] ^0Miles]]]. See the decomposition, with full typing, in Duží et al. (2010: 78). The decomposition in Duží et al. (2010a: §3) left out one part: [λt [[[^0Happy w] t] $^0Tadeusz$]].

property of having nine sub-constructions (i.e. proper parts) of it. We can go on to predicate of it what its unifiers are. To do so we must ascend to one order higher than the order of these unifying procedures and from that vantage point explain via predication why this multi-entity hyperproposition is a unit. This ascent is a matter of expressive power, and the required power derives from the ramified type hierarchy of higher-order entities that is part and parcel of TIL.[26]

The default mode in which constructions occur is the *executed* mode. This stipulation renders superfluous the specification that a given construction is to be executed.[27] This explains why the fifth construction known as *Single Execution* is not required to explicitly convert a construction from displayed to executed mode. (Above when the number of parts ran to nine the implicit instances of Single Execution were not included.) Where a displayed construction *describes* its parts and their interconnection by showing rather than telling, Single Execution *prescribes* that the construction it operates on be executed.[28]

4. Discussion and Conclusion

This paper poses the question whether predication is an act or an operation. The answer is that we can have it both ways. Predication can be both an act and an operation. But the same notion of predication cannot be both. So there is predication as an act, as something people do, and there is predication as an operation, as a logical nexus between (the modally and temporally qualified extension of) a property and an individual.

The difference can be characterized as a type-theoretic difference. Semantic or objective predication is an instance of the procedure of functional application. Predication is the procedure *describing* the application of an extensionalized property (i.e. the characteristic function of a set) to an individual in order to obtain a truth-value. The procedure displays how different logical objects (functions and arguments) are *connected inter se*. Derivatively, the procedures are also

---

26 See Duží et al. (2010: 52–53).
27 $^1[\lambda w \lambda t\ [^0Smiles_{wt}\ ^0Miles]]$ is the Single Execution of the Closure $\lambda w \lambda t\ [^0Smiles_{wt}\ ^0Miles]$. In the light of the default stipulation that constructions occur in executed mode, these constructions are one and the same construction. Also note that Single Execution distributes, such that each sub-construction within the Closure is affected by Single Execution.
28 My (ms.) explains in detail how Composition and Closure serve as propositional unifiers.

connected inter se: how they are interconnected is a reflection of how the entities they produce are interconnected. Whenever we apply the procedure of Single Execution to the displayed procedure, we thereby heed the instruction to execute predication. Whether the instruction is in fact heeded by an agent is an empirical and contingent matter that is beyond the nature and identity of the procedures.

At no point so far has anyone done anything. As soon as agents enter the fray, they must be added as additional functional arguments. Therefore, pragmatic or subjective predication has got to be a type-theoretically distinct mapping.[29] In TIL pragmatic predication comes out a three-argument function that models a relation (-in-intension) from the individual who is the subject of the predicative act, the property that is predicated, and the individual of whom or which it is predicated, to a truth-value. This truth-value is in turn indexed to the empirical parameters (worlds and times) at which the predicative act takes place. Accordingly a predicative act is the sort of act agents perform when they apply an extensionalized property to an individual. I am keeping it simple by assuming that the predicative act takes place in the same world and at the same time that the attributee is predicated to instantiate the property in question.

We could turn the conceptual priorities around, such that the operation of predication would be tracking the act of predication. This would presumably involve the sort of correctness conditions for acts that Hanks (2015: Ch. 3) invokes. But conditions that sort acts into those that are successful and those that are failures still do not answer the question what the *logical* nexus is between a property and an individual when the former is predicated of the latter. When objective predication is conceptually prior to subjective predication, we can point to the instruction

---

29 For reference, the TIL construction is $\lambda w \lambda t \ [^0Predicate_{wt} \ ^0a \ ^0F \ ^0b \ ]$, where *Predicate* of type $(o\iota(o\iota)_{\tau\omega}\iota)_{\tau\omega}$ is a relation-in-intension which, at a world and a time, takes a triple consisting of an individual of type ι (the attributer), a property of type $(o\iota)_{\tau\omega}$ (the property being pragmatically predicated), and an individual (the attributee) to a truth-value of type o. *Predicate*, thus typed, is a far cry from Hanks's notion of predication, which I propose typing as a function from an individual and a property to a truth-condition when the proposition is $\vdash \langle a, F \rangle$. I presume we may parse "$\vdash \langle a, F \rangle$" as 'Predicate F of a' or 'Assert F of a' and "$\sim \vdash \langle a, F \rangle$" as 'In a cancellation context, predicate F of a', though not as 'In a cancellation context, assert F of a'. Hanks's general symbolism takes existing notation and reinterprets it act-theoretically. This is in itself unobjectionable, but it is not yet clear how to draw inferences from his propositions as premises, nor is it entirely settled which combinations of symbols are well-formed and meaningful and which are not. This is because the symbolism comes without formation rules at its present stage.

to apply a function as the logic of objective predication and as the mechanism that is invoked when agents perform subjective acts of predication.

Acknowledgments

The research reported herein was supported by Marie Curie Fellowship FP-7-PEOPLE-2013-IEF 628170 USHP and by the Grant Agency of the Czech Republic Project GA15–13277S, *Hyperintensional Logic for Natural Language Analysis*. I am indebted to Marie Duží for precious comments.

References

Duží, M., Jespersen, B. and Materna, P. (2010). *Procedural Semantics for Hyperintensional Logic. Foundations and Applications of Transparent Intensional Logic*, LEUS, vol. 17, Berlin: Springer.

Duží M., Jespersen, B. Materna, P. (2010a). "The *Logos* of Semantic Structure". In: P. Stalmaszczyk (ed.), *Philosophy of Language and Linguistics*, vol. 1: *The Formal Turn*, Frankfurt et al.: Ontos-Verlag, 85-101.

García-Carpintero, M. (2010). "Gaskin's Ideal Unity". *Dialectica* 64: 279–288.

Gaskin, R. (2008). *Unity of the Proposition*. Oxford: Oxford University Press.

Gilmore, C. (2014). "Parts of Propositions". In: S. Kleinschmidt (ed.), *Mereology and Location*, Oxford: Oxford University Press, 156–208.

Hanks, P. (2015). *Propositional Content*. Oxford: Oxford University Press.

Jespersen, B. (2003). "Why the Tuple Theory of Structured Propositions Isn't a Theory of Structured Propositions". *Philosophia* 31: 171–183.

Jespersen, B. (2008). "Predication and Extensionalization". *Journal of Philosophical Logic* 37: 479–499.

Jespersen, B. (2012). "Recent Work on Structured Meaning and Propositional Unity". *Philosophy Compass* 7: 620–630.

Jespersen, B. (2012a). "Post-Fregean Thoughts on Propositional Unity". J. Maclaurin (ed.), *Rationis Defensor. Essays in Honour of Colin Cheyne, Studies in the History and Philosophy of Science*, vol. 28, 235–254.

Jespersen, B. (2015). "Should Propositions Proliferate?". *Thought* 4: 243–251.

Jespersen, B. (ms.). "Anatomy of a Proposition", paper in submission.

Johnson, M. (2006). "Hylomorphism". *Journal of Philosophy* 103: 652–698.

King, J. (2007). *The Nature and Structure of Content*. Oxford: Oxford University Press.

King, J. C., Soames, S. and Speaks, J. (2014). *New Thinking About Propositions*. Oxford: Oxford University Press.

Lenci, A. (1998). "The Structure of Predication". *Synthese* 114: 233–276.

Levine, J. (2002). "Analysis and Decomposition in Frege and Russell". *Philosophical Quarterly* 52: 195–216.

Liebesman, D. (2015). "Predication as Ascription". *Mind* 124: 517–569.

Merricks, T. (2015). *Propositions*. Oxford: Oxford University Press.

Reiland, I. (2012). "Propositional Attitudes and Mental Acts". *Thought* 1: 239–245.

Russell, B. (1910). "On the Nature of Truth and Falsehood". In: *Philosophical Essays*, London: Longmans, Green & Co., 147–159.

Russell, B. (1903). *Principles of Mathematics*. New York: Norton Library.

Soames, S. (2010). *What Is Meaning?* Princeton: Princeton University Press.

Tichý, P. (1995). "Constructions as the Subject-matter of Mathematics". Reprinted in C. Cheyne, B. Jespersen, V. Svoboda (eds.), *Pavel Tichý's Collected Papers in Logic and Philosophy*, University of Otago Press and Filozofia, Academy of Sciences of the Czech Republic (2004), 873–885.

Williams, E. (1980). "Predication". *Linguistic Inquiry* 11: 203–238.

Jacek Paśniczek
Maria Curie-Skłodowska University, Lublin, Poland

Meinongian Predication. An Algebraic Approach

Abstract: Alexius Meinong, an Austrian philosopher, is particularly famous for his abundant ontology embracing various types of non-existent objects. The most distinguished formal-ontological feature of his theory of objects is that it allows for impossible and incomplete objects like *the round square*. Recently there have appeared several logical interpretations of Meinong's theory of objects rendering it consistent. Generally, in order to render such a theory consistent we have to modify the classical concept of predication according to which it is a relation holding between objects and properties, i.e. basic formulas of respective formal languages are subject-predicate ones. In the present paper we are going to develop an unorthodox view of Meinongian predication based on a calculus of names. The calculus is semantically interpreted in an algebraic structure which is an extension of De Morgan lattice (the De Morgan negation as basically weaker than the Boolean negation enable us to interpret impossible objects in a consistent way). Roughly speaking, the idea is that predication is expressed by a formula '*a is b*' where *a* and *b* are of the same syntactic category, i.e. the name category. This means that the roles of properties in the predication are played by objects of Meinongian kind. Within the proposed algebraic setting various ontological notions pertaining Meinong's theory of objects are studied. Let us mention here notions of contradictory and complementary properties, notions of consistency, completeness and individuality of objects. Most ontological issues considered in the paper are almost untouched in the contemporary discussions concerning formal aspects of Meinong's theory of objects. Some formal results achieved in the paper are quite important for ontology and philosophy of language and, sometimes, are a little bit surprising. In particular, it turns out that consistent and complete objects need not be individual ones as it is usually assumed.

Keywords: atoms of Meinongian lattice, complementary properties, contradictory properties, complete lattice, De Morgan lattice, De Morgan negation, existence, filtration function, greatest element, (in)complete objects (M-objects), (in)consistent objects (M-objects), individuals (individual M-objects), least element, M-objects (Meinongian objects), Meinong's theory of objects, Meinongian lattice, atomic Meinongian lattice, set-theoretical meinongian lattice, predication, Meinongian predication, universals (universal M-objects)

0. Introduction. Incompleteness and Inconsistency of Meinongian Objects

Almost every formal theory or logic of nonexistent objects refers to Meinong, a famous Austrian philosopher living at the turn of the 19th and 20th centuries. But the mere possibility of objects' nonexistence is perhaps not the most distinctive feature of Meinong's ontology. What distinguishes this ontology from others is that objects not only may not exist but they may be incomplete or inconsistent.[1] Although incompleteness and inconsistency are pivotal notions of Meinong's theory of objects, not enough attention has been paid to them so far. Usually, to explain the meanings of these notions, simple examples of incomplete and inconsistent objects have been given without any reference to a more general theoretical explanation. Thus, *the round square* (or: *the non-squared square*) is inconsistent and incomplete since it possesses two contradictory properties of *roundness* and squareness; it is incomplete since it possesses neither the property of *redness* nor *non-redness*. Generally, an object is incomplete if for some property *P* it possesses neither *P* nor the complementary property *non-P*; an object is inconsistent if it possesses contradictory properties. But how do we know that there is such an object as *the round square*? This is guaranteed by the so called principle of unrestricted free assumption or *Annahmen* thesis which says that every set (class) of properties constitutes exactly one object.

Clearly, the property negation *non* cannot be reducible to propositional negation because talking about inconsistent objects would then result in a plain contradiction; *the non-squared square* will at the same time possess and not possess the property *squareness*. Expressing this formally, if *a* means *the non-squared square* and *P* means *is a square* then we will get: Pa and $\neg Pa$.

It seems quite obvious that the notions of inconsistency and incompleteness of objects are strictly based on the notions of contradictory and complementary properties. But how can the latter be defined? Let us begin with the notion of complementarity. Certainly, for any property *P* the complementary property *non-P* cannot be defined as:

(*) $\forall x \, (non\text{-}Px \equiv \neg Px)$

---

[1] In this paper I use the term 'inconsistent' instead of the more often used word 'impossible'. This is because I usually associate with the term 'inconsistent' a technical meaning. Also, I would like to avoid any associations of the term 'impossible' with modality. Notice that when we talk about nonexistent objects we usually have in mind nonexistent individuals, i.e. complete and consistent objects existing in some other possible world.

with an all-embracing domain of quantification since such a definition would rule out inconsistent and incomplete objects. In other words, no object could possess both *P* and *non-P*. For another reason, one could hardly accept the narrowing of the range of the quantifier in the definition (*) to existent objects only. Such a solution would make the notion of property complementation dependent on what there is, i.e. on the domain of existing objects. We are obviously not willing to do so: (in)consistency and (in)completeness as formal-ontological notions should not be dependent on the metaphysical notion of existence. In particular, completeness and consistency do not and cannot entail existence. On the other hand, however, it is a commonly shared view among metaphysicians, and in particular among Meinong commentators, that all existing objects – existing individuals – are consistent and complete.

There is still another way of defining the complement of property by suitable adoption of the domain of quantification in (*): we may allow all possible individuals to be values of variables. Merely possible individuals are still complete and possible objects but they do not need to exist. Positing the domain of possible individuals is not quite neutral ontologically and leads to a kind of modal actualism. But certainly, being committed to possible individuals seems to be more neutral with respect to theoretical generality of completeness and consistency.

What about contradictory properties such as *roundness* and *squareness*? We simply know that the two properties are contradictory. But how can we generally discover that two given properties, say *P* and *Q*, are contradictory? In the case of complementarity, this feature of properties was formally expressed by a kind of negation *non*. Needless to say, the contradiction of *P* and *Q* can be expressed in the following way: *P* entails *non-Q* or *Q* entails *non-P*. But what does it mean that one property *entails* another?[2] The relation of entailment between properties may be expressed by means of the following formulas:

(**) $\forall x(Px \supset \neg Qx)$ or $\forall x(Qx \supset \neg Px)$

But then again we face the problem of the domain of quantification. We cannot allow inconsistent objects to be the value of quantification since in particular

---

2   Here I adopt frequently used terminology even though it is not quite adequate. The complement of property *P*, *non-P* is the weakest contradictory property in the sense of the given entailment. The hardly discussed dual to the notion of contradictory properties is something like a *weak complementarity*. It is difficult to find a natural example of a pair of weakly complementary properties besides those involving property negation, e.g. *rectangle, non-square*. From this point of view, *non-P* is the strongest weakly complementary property of *P* in the sense of entailment.

properties like *roundness* and *squareness* would then turn out to be noncontradictory as possessed by at least one inconsistent object. If the domain contains only existing objects, then such an entailment will be completely trivial for it is tantamount to the disjointness of extensions of *P* and *Q*. As in the case of previously discussed notions, more acceptable, although not uncontroversial, is the assumption that the domain of quantification in (**) is equal to the class of possible individuals.

There is, however, a way of avoiding any ontological commitments while introducing the notion of property complement (or property negation). We can treat the complement as a primitive, non-definable operation which when applied to a property *P* yields another property *non-P*. This corresponds to the syntactical operation of prefixing a predicate by the word *non*. At the same time, we generally do not accept (*).[3] The meaning of *non*, however, is far from being unambiguous: if we do not explicitly declare what meaning we have in mind, then the Boolean negation will be assumed by default. But, as it is well known, the Boolean negation involves the maximal element, which in case of Boolean algebra of sets is just the universal set. So, in our case, if *non* is to be the Boolean negation, we will have to introduce a unique maximal property i.e. the property possessed by every object. Thus, here again we face the problem discussed earlier. This unwelcome consequence has been hardly ever realized by philosophers dealing with Meinongian objects.

As our remarks have shown, it is not so easy to make the notions of (in)complete and (in)consistent objects, as well as the related concepts of property complement and property entailment, theoretically well grounded. In what follows, we will approach these notions within an algebraic setting starting with a negation weaker than the Boolean one, i.e. the De Morgan negation (which is applied in various non-classical logics).[4] This negation will not directly render the meaning of *non*, although somewhat surprisingly, it will indirectly provide an interpretation of (in)completeness and (in)consistency of objects. We are going to develop an algebraic structure on the ground of which we can study various formal aspects of Meinong's theory of objects including logical relations between the notions mentioned above. This algebraic structure will be a kind of De Morgan lattice equipped with an additional operation of filtration. This lattice will be called Meinongian lattice.

---

3 The bivalence is only true for individual objects which, as we have noticed above, are complete and consistent.
4 See e.g. Dunn and Hardegree (2001), section 3.13.

Since (in)completeness and (in)consistency of Meinongian objects pose serious logical problems, they provoke one to look for a consistent rendering. Recently, several formal theories of non-existent objects have been put forward.[5] They are more or less faithful interpretations of Meinong's original views or are merely inspired by them. All these theories are based on the classical calculus of predicates and they are often equipped with an additional strong logical machinery including second order logic devices. In the simplest case, the Meinongian predication takes the form of subject-predicate formulas Pa in which the subject a and predicate P are of different syntactic and semantic categories. Roughly, a refers to a Meinongian object $I(a)$ which is characterized by or identified with a set of properties and the interpretation of P is a single property $I(P)$. Formula Pa is true if and only if $I(P) \in I(a)$. This is perhaps the most straightforward semantical approach to Meinongian predication.[6]

However, our goal here is different. We are going to render the Meinongian predication within a calculus of names interpreted in the aforementioned algebraic structure.[7] Such an interpretation presupposes only one syntactic and semantic category instead of two categories, i.e. categories of names and predicates on the one hand, and categories of objects and properties on the other. Thus, the formula '*a is b*' will express the Meinongian predication where a and b will represent Meinongian objects. The objective of this paper is to find an algebraic relation for the interpretation of *is*.

1. Meinongian Predication in De Morgan Lattice

One can wonder how it is possible to talk about predication and properties within the language of name calculus. Here, we offer the reader a hint. Roughly speaking, objects referred to by b in '*a is b*' will play the role of properties or universals which are predicated of objects referred to by a. Let us consider the following sentences and discuss their truth values:

---

[5] Cf. Jacquette (1996), Paśniczek (1998), Parsons (1980), Routley (1980), Zalta (1983), Zalta (1988), Zalta (1999). Cf. also some general approaches to Meinong's theory of objects: Berto (2013), Jacquette (2015), Perszyk (1993), Swanson (2011).

[6] However, it should be stressed that in the formal theories inspired by Meinong's views the classical predication associated with the calculus of predicates is modified in one or another way in order to avoid inconsistencies. In particular, the theories adopt the conception of two kinds od of properties (Parsons, Jacquette, Routley) or the conception of two kinds of predication (Paśniczek, Zalta). Here we are dealing only with "encoding" (Zalta) or internal predication (Paśniczek).

[7] Cf. my earlier discussion of this issue: Paśniczek (1996), Paśniczek (2014b).

(I)   The round square is round
(II)   The round square is a square
(III)   The round square is a geometrical figure
(IV)   The round square is non-round
(V)   The round square is not a square
(VI)   The round square is a round square
(VII)   The round square is black
(VIII) The round square is non-black

If we understand *is* as the inclusion relation between the extensions of subjects and predicates of the sentences (I)–(VIII), then the sentences will turn out to be true for the obvious reason that the first extension is vacuous. When we interpret the subject as definite description, then sentences (I)–(VIII) will turn out to be false. Also, all of the sentences will be false if we interpret them as expressing a formula $a \varepsilon b$ within the Leśniewskian ontology. We can say that the three classical approaches to the truth value of (I)–(VIII) make any predication involving non-existent and – in particular – inconsistent objects, trivial. The situation will not become much better if we appeal to modal logics.

Most of us will perhaps recognize every predication of the form

(***)   'pq is p' and 'pq is q'[8]

as analytic. This simple intuition also underlies Meinongially oriented logics – in all of them sentences (I) and (II) are true. But what makes the approach to non-existent objects non-trivial is that at the same time these logics agree that the last two sentences (VII) and (VIII) are false, which means that the object *the round square* is an incomplete object. The logics differ only with respect to whether Meinongian objects are allowed to be deductively closed in some sense, which determines whether they treat (III)–(VI) as true or false.

To get a better understanding of Meinongian predication, let us propose a naive formal model. Let us assume that Π is some fixed class of properties. According to Meinong's theory of objects, for every subclass P of Π there is an object possessing these and only these properties which are members of P (*Annahmen* thesis). This does not mean that we claim that Meinong's theory of objects must be a strong bundle theory, i.e. a theory in which objects are identified with sets of properties. We assume only a weaker version of the bundle theory according to which there is a one-to-one correspondence between objects and sets of properties.[9] For example, let $\Pi=\{p,q,r\}$. Then we will have the following objects: $\emptyset$, $\{p\}$, $\{q\}$, $\{r\}$,

---

8   Roughly, *p and q* are general terms and "*pq*" is their concatenation.
9   To make things easier we will often directly represent objects by sets of properties.

{p,q}, {p,r}, {q,r}, {p,q,r}.[10] Objects possess exactly those properties which are members of respective sets (e.g. {p,r} possesses p and r). Which objects, however, are complete, incomplete, consistent, inconsistent? When only this purely theoretical setting is given, we generally cannot answer these questions. We have to know first which pairs of properties are contradictory and which are complementary. It is possible that there are no contradictory or complementary properties among {p,q,r}. We can only say that, for example, if {p,q} is consistent then so must be {p} and {q}; if {p,q} is inconsistent then so must be {p,q,r}; if {p,q} and {p,r} are consistent then so must be {p}; if {p,q} is complete then so must be {p,q,r}; if {p,q} is incomplete then so must be {p} etc.

In general, given a class of properties Π, the power set $\wp(\Pi)$ of Π will represent the class of objects. These objects are partially ordered by the inclusion relation $\subseteq$: $a \subseteq b$ means that every a's property is b's property. Therefore, the converse of relation $\subseteq$, i.e. the relation $\supseteq$ may express predications of the form (***). The formula $b \supseteq a$ can be read: b is a and this is our preliminary notion of Meinongian predication. An object a possesses a property p iff $p \in a$, i.e. iff $\{p\} \subseteq a$. But we can also say that $\{p,q,r\}$ is $\{p,q\}$ in the sense that $\{p,q\} \subseteq \{p,q,r\}$. From the algebraic point of view, the structure $\langle \wp(\Pi), \subseteq \rangle$ is a lattice. This easy observation may be generalized – we may assume that the ontology of objects is represented by a structure $\langle M, \leq \rangle$, where M is the class of Meinongian objects and $\leq$ is their partial ordering. Alternatively, the lattice $\langle M, \leq \rangle$ can by displayed in the form $\langle M, \sqcup, \sqcap \rangle$, where $\sqcup$, $\sqcap$ are join and meet operations respectively. $a \sqcup b$ is to be an object that possesses every property that is possessed by a or by b; $a \sqcap b$ is to be an object that possesses every property that is possessed by a and by b. In many cases we will still refer to objects as sets of properties, mostly for the sake of simplicity.

Needless to say, this formal approach to Meinongian objects is almost completely trivial; in particular, we cannot distinguish (in)complete and (in)consistent objects. So let us try to enrich the lattice structure with an additional operation called De Morgan negation ¯. The operation is characterized by the following conditions:

DM1 $\quad \overline{\overline{a}} = a$

DM2 $\quad a \leq b \equiv \overline{b} \leq \overline{a}$

DM3 $\quad \overline{a \sqcap b} = \overline{a} \sqcup \overline{b}$

DM4 $\quad \overline{a \sqcup b} = \overline{a} \sqcap \overline{b}$[11]

---

10 $\emptyset$ is not always allowed to represent an object.
11 De Morgan laws **DM3** and **DM4** follow from **DM1** and **DM2**.

It is worth emphasizing that equalities $a \sqcup \bar{a} = V$ and $a \sqcap \bar{a} = \emptyset$ will not generally hold even if the lattice has the maximum and the minimum elements V and $\emptyset$. A distributive lattice $\langle M, \sqcup, \sqcap, ^- \rangle$ with De Morgan negation is called De Morgan lattice. What is particularly important for us, is that in the De Morgan lattice the following may be true: $a \leq \bar{a}$, $\bar{a} \leq a$, and $a = \bar{\bar{a}}$.[12] The De Morgan negation occurring in the formulas can hardly be read as "not." How could we understand: "a is not a" or "not a is a"? However, there will be a very important meaning associated with these three formulas. To see this, let us consider the following, easy to prove theorems of De Morgan lattice:

(1) $b \leq a \wedge a \leq \bar{a} \supset b \leq \bar{b}$

(2) $a \leq \bar{a} \wedge b \leq \bar{b} \supset a \sqcap b \leq \overline{a \sqcap b}$

(3) $b \leq a \wedge \bar{b} \leq b \supset \bar{a} \leq a$

(4) $\bar{a} \leq a \wedge \bar{b} \leq b \supset \overline{a \sqcup b} \leq a \sqcup b$

Assume provisionally that $a \leq \bar{a}$ means that a is consistent and $\bar{a} \leq a$ means that a is complete. The theorems can be now read in the following way, respectively: (1) if a is b (i.e. every property of b is a property of a) and a is consistent then b must be consistent; (2) if a and b are consistent then their meet must be consistent; (3) if a is b (i.e. every property of b is a property of a) and b is complete then a must be complete as well; (4) if a and b are complete then their join must be complete as well. Hence such a reading of $a \leq \bar{a}$ and $\bar{a} \leq a$ will perfectly agree with the classical meaning of consistent and complete sets of formulas as long as symbols $\leq$, $\sqcap$, and $\sqcup$ are understood set-theoretically (as inclusion, intersection, and union respectively). So, let us define formally consistency and completeness:

df-con$_0$ $\text{con}_0(a) \equiv_{df} a \leq \bar{a}$

df-com$_0$ $\text{com}_0(a) \equiv_{df} \bar{a} \leq a$

Notice that a is consistent iff $\bar{a}$ is complete and *vice versa* ($\text{con}_0(a)$ iff $\text{com}_0(\bar{a})$ and $\text{con}_0(\bar{a})$ iff $\text{com}_0(a)$). Still, it is not easy to recognize the proper meaning of De Morgan negation and consequently, we have no guarantee that con_0 and com_0 express the common-sense meaning of consistency and completeness. In what follows, we will enrich De Morgan lattice with an additional operation which will enable us to introduce more advanced ontological concepts, in particular, that of Meinongian predication.

---

12 These are not true in Boolean algebra.

It is worth noticing that according to the present approach, since objects may be constituted of more than one property, the second argument of predication may consist of two or more properties like in the following example: $\{p,q,r\}$ is $\{p,q\}$. This makes our predication unusual when compared with the classical predication in which the second argument is always a single property. Besides, our way of talking about Meinongian objects as consisting of properties is merely heuristic. Officially, we should talk only about objects and, in particular, about these objects which are the second arguments of predication (i.e. objects representing single properties). Yet, unofficially, in order to make our considerations more understandable, we will continue to refer to objects as constituted of properties (or simply, as sets of properties).

2. Meinongian Predication in Meinongian Lattice

Now, we extend De Morgan lattice $\langle M, \sqcup, \sqcap, ^- \rangle$ to *Meinongian* lattice $\mathcal{M} = \langle M, \sqcup, \sqcap, ^-, f \rangle$, where elements of lattice are interpreted as Meinongian objects (in short: M-objects), f is an operation called *filtration function* defined on M and fulfilling the following conditions:

$\mathcal{M}5$[13] $\quad a \leq f(a)$

$\mathcal{M}6 \quad\quad f(a \sqcap b) = f(a) \sqcap f(b)$

$\mathcal{M}7 \quad\quad f(f(a)) = f(a)$

Intuitively, f is a function which 'filtrates' a set of properties i.e. assigns to the set of properties the set consisting of a single property. So, if a is an M-object constituted of number of properties then $f(a)$ will be M-object constituted of just one property which can be understood as the conjunction of properties of a. E.g., to the M-object {being red, being a car} f assigns the M-object {being a red car}.

Now, we define the most important notion of our algebraic approach, i.e. the notion of Meinongian predication:

Df $\in \quad\quad a \sqcup c \in b \equiv_{df} f(b) \leq a \vee f(b) \leq c$

However, in what follows, we will usually refer to a weaker version of this definition (we just put a for c):

df $\in \quad\quad a \in b \equiv_{df} f(b) \leq a$

---

[13] In order to distinguish axioms from theses of De Morgan lattice we use italics for the former and bold for the latter.

Notice that this relation of predication is stronger than the converse of the relation of ordering for M-objects, i.e. the relation $\geq$, which was introduced as a preliminary notion of Meinongian predication (if $f(b) \leq a$ then according to $\mathcal{M}5$, also $b \leq a$). The difference between the two types of predication as a formal explication of informal predication is such that now the second argument (i.e. M-object $f(b)$) represents a single property.

Let us list some theorems which hold for $\mathcal{M}^{14}$:

M8 $a \Subset a \equiv a = f(a)$

M9 $a \Subset b \wedge b \Subset a \supset a = b$

M10 $a \Subset b \wedge b \Subset c \supset a \Subset c$

M11 $a \leq b \supset f(a) \leq f(b)$

M12 $a \Subset b \supset b \leq a$

M13 $a \Subset b \wedge c \leq b \supset a \Subset c$

M14 $\overline{f(a)} \leq f(\overline{a})$

M15 $a \Subset b \equiv a \Subset f(b); f(a) \Subset f(a)$

M16 $a \leq \overline{a} \supset (a \Subset b \supset b \leq \overline{b})$

M17 $f(a \sqcup b) = f(f(a) \sqcup f(b))$

M18 $a \sqcup c \Subset b \equiv a \Subset b \vee c \Subset b$

Now, let us define the notions of *universal element* and *atom* of $\mathcal{M}$:

df $\mathcal{U}$ $\mathcal{U}(a) \equiv_{df} a \Subset a$
df $\mathcal{A}$ $\mathcal{A}(a) \equiv_{df} a = \overline{a} \wedge a \Subset a$[15]

Since universal M-objects are fixed points of filtration f (see **M8**), we will often interpret the objects as properties and simply call them *properties*. Clearly, every

---

14 Most proofs of the theorems are pretty easy, some of them are complicated, but rather elementary, cf. Paśniczek (2014a).
15 $\mathcal{U}$ and $\mathcal{A}$ will be used not only as predicate symbols, but also as set symbols for the set of all universal elements and atoms of the lattice respectively. Notice that, in particular, $\mathcal{A} \subseteq \mathcal{U}$.

Meinongian Predication. An Algebraic Approach 257

M-object $f(a)$ is a universal M-object (cf. **M15**). By definition **df** $\mathcal{A}$, atoms are consistent ($a \leq \bar{a}$) and complete ($\bar{a} \leq a$); moreover, they exist in some reasonable sense of existence which will be defined below. Therefore, we may interpret them as *individual M-objects* or simply *individuals*.

We can define an important notion of non-contradictory M-objects (properties):

df-noncontrad$_0$ noncontrad$_0(a,b) \equiv_{df} \bar{a} \in b$

Although this definition applies to all M-objects, its meaning concerns mainly universal M-objects. The following theorem shows that no consistent M-object, in the sense of **con**$_0$, possesses a pair of contradictory properties:

M19 $a \leq \bar{a} \supset (a \in b \land a \in c \supset \bar{b} \in c)$

As we will see, in a stronger version of Meinongian lattice the reverse implication also holds (see **M**$^+$**34**). Unfortunately, though the notions of contradictory and (weakly-) complementary properties are in a sense dual, there is no definition of the second notion which could be in a way analogous to that of **noncontrad**$_0$.

The following theorems characterize the notion of atom:

M20 $\mathcal{A}(a) \equiv a = \bar{a} \land a = f(a)$

M21 $\mathcal{A}(a) \equiv a \in a \land \bar{a} \in \bar{a}$

M22 $\mathcal{A}(c) \land c \in a \land c \in \bar{a} \supset \mathcal{A}(a)$

M23 $\mathcal{A}(a) \land \mathcal{A}(b) \supset (a \in b \supset a = b)$

M24 $\mathcal{A}(c) \land c \in a \supset a \leq \bar{a}$

M25 $\mathcal{A}(c) \land a \in c \supset \bar{a} \leq a$

M26 $\mathcal{A}(c) \supset (c \in a \equiv \bar{a} \in c)$

M27 $\mathcal{A}(c) \supset (c \in a \sqcup b \equiv c \in a \land c \in b)$

The version of Meinongian lattice developed so far is relatively poor and, in particular, it is quantifier free (i.e. every use of quantifiers is an inessential one). A stronger version of Meinongian lattice that meets some details of Meinongian theory of objects should be considered now. First of all, we assume that our Meinongian lattice is complete, i.e. for every $X \subseteq M$ there exist infimum ($\sqcap X$) and supremum ($\sqcup X$):

df$\sqcap$ $a = \sqcap X$ iff for all $b \in X$, $a \leq b$, and if for all $b \in X$, $c \leq b$, then $c \leq a$,

df⊔ $a = \sqcup X$ iff for all $b \in X$, $b \leq a$, and if for all $b \in X$, $b \leq c$, then $a \leq c$[16]

The completeness means that the lattice has the *greatest* element $\Omega = \sqcup M$ for which for all b, $b \leq \Omega$ and the *least* element $\mho = \sqcap M$ for which for all b, $\mho \leq b$.

The following theorem is a generalization of **M27**:

M28 $\mathcal{A}(c) \supset (c \Subset \sqcup X \equiv \forall d \in X (c \Subset d))$

3. Stronger Meinongian Lattices

The Meinongian complete lattice $\mathcal{M}$ is now enriched with the additional axiom $\mathcal{M}^+29$. The resulting lattice will be called $\mathcal{M}^+$ or atomic Meinongian lattice.[17]

$\mathcal{M}^+29$ $\forall c(a \Subset c \supset b \Subset c) \supset a \leq b$

The following theorems can be proved in $\mathcal{M}^+$:

M⁺30 $a \leq b \supset \forall c(a \Subset c \supset b \Subset c)$

M⁺31 $a = b \equiv \forall c(a \Subset c \equiv b \Subset c)$

M⁺32 $a \leq b \equiv \forall c\, (\mathcal{U}(c) \supset (a \Subset c \supset b \Subset c))$

M⁺33 $a = b \equiv \forall c(\mathcal{U}(c) \supset (a \Subset c \equiv b \Subset c))$

M⁺34 $\forall b,c\, (a \Subset b \wedge a \Subset c \supset \overline{b} \Subset c) \supset a \leq \overline{a}$

The Meinongian lattice $\mathcal{M}^+$ can be extended to an even stronger one – the atomic Meinongian lattice $\mathcal{M}^{++}$ – by adding the following axiom $\mathcal{M}^{++}35$:

$\mathcal{M}^{++}35$ $f(b) \leq \overline{f(c)} \supset \exists a\, (\mathcal{A}(a) \wedge a \Subset b \wedge a \Subset c)$

Let us list theorems which characterize more closely the atomicity of Meinongian lattice $\mathcal{M}^{++}$.

M⁺⁺36 $f(b) \leq \overline{f(b)} \supset \exists a\, (\mathcal{A}(a) \wedge a \Subset b)$

M⁺⁺37 $\mathcal{U}(a) \wedge \mathcal{U}(b) \supset (a \leq b \equiv \forall c(\mathcal{A}(c) \supset (c \Subset b \supset c \Subset a)))$

---

16 Roughly speaking, $\sqcap X$ is the result of applying the meet operation to possibly infinite subset of M; $\sqcup X$ is the result of applying the join operation to possibly infinite subset of M.

17 In this lattice the use of quantifiers will be an essential one.

M⁺⁺38 $\mathcal{U}(a) \wedge \mathcal{U}(b) \supset (a = b \equiv \forall c(\mathcal{A}(c) \supset (c \in b \equiv c \in a)))$

M⁺⁺39 $f(b) \leq \overline{f(b)} \supset \exists a\,(\mathcal{A}(a) \supset a \in b)$

M⁺⁺40 $\mathcal{A}(a) \equiv f(\overline{a}) = \overline{f(a)}$

M⁺⁺41 $\mathcal{U}(a) \wedge \mathcal{U}(b) \supset (a \leq b \equiv \forall c(\mathcal{A}(c) \supset (c \in b \supset c \in a)))$

M⁺⁺42 $\mathcal{U}(a) \wedge \mathcal{U}(b) \supset (a = b \equiv \forall c(\mathcal{A}(c) \supset (c \in b \equiv c \in a)))$

M⁺⁺43 $a \leq \overline{a} \supset \mathcal{U}(a) \supset \exists c(\mathcal{A}(c) \wedge c \in a))$

M⁺⁺44 $\exists a(a \leq \overline{a} \wedge a \in b) \equiv \exists a(\mathcal{A}(a) \wedge a \in b)$

M⁺⁺45 $\neg \forall b\,(a \in b) \supset \exists c\,(\mathcal{A}(c) \wedge c \in a)$

M⁺⁺46 $\neg \forall b\,(a \in b) \supset (\mathcal{U}(a) \supset a \leq \overline{a})$

M⁺⁺47 $\mathcal{A}(c) \supset (c \in a \sqcap b \equiv c \in a \vee c \in b)$

M⁺⁺48 $\mathcal{A}(c) \supset (c \in \sqcap X \equiv \exists d\varepsilon X\,(c \in X))$

M⁺⁺49 $f(a) = a \supset (a \in f(b) \equiv \forall c\,(b \in c \supset a \in c))$

The following two theorems describe the canonical structure of M-objects in $\mathcal{M}^{++}$:

M⁺⁺50 $\forall a \exists X\,(X \subseteq \mathcal{U} \wedge a = \sqcup X)$

M⁺⁺51 $\forall a(\mathcal{U}(a) \supset \exists X\,(X \subseteq \mathcal{A} \supset a = \sqcap X)$

Proof of **M⁺⁺50**
Let $X = \{b : a \in b\}$. Suppose that $\neg a \leq \sqcup X$. By **M⁺32**, there exists c, $\mathcal{U}(c)$, $a \in c$, and $\neg \sqcup X \in c$. Thus $c \in X$ and what follows $c \leq \sqcup X$. Hence $\sqcup X \in c$. Contradiction.
Assume now that for every $b \in X$, $a \in b$. Hence $b \leq a$ by **M12** and according to **df⊔**, $\sqcup X \leq a$. ∎

Proof of **M⁺⁺51**.
Let $U(a)$ and $X = \{b : \mathcal{A}(b) \wedge b \in a\}$. Suppose that $\neg \sqcap X \leq a$. Since $\sqcap X \in \sqcap X$ then by **M⁺32** there exists c, $\mathcal{A}(c)$, $c \in a$, and $\neg c \in \sqcap X$. What follows $c \in X$ and $\sqcap X \leq c$. Thus $c \in \sqcap X$. Contradiction.

Assume now that for every $b \in X$, $b \sqsubseteq a$. Hence $a \leq b$ by **M12**, and according to **df⊓**, $a \leq \sqcap X$. ∎

M⁺50 and **M⁺51** say that every element of $\mathcal{M}^+$ can be displayed in the following canonical form: $\sqcup_{i \in I} \sqcap_{j \in J} a_{ij}$[18] where a_{ij} are atoms. Since $\sqcap_{j \in J} a_{ij}$ is a universal element for every $i \in I$ (see $\mathcal{M}$ 6), every element of $\mathcal{M}^+$ is a join of universal objects. This formal representation expresses the fact that M-objects are "bundles of properties."

Since the lattice is distributive, the canonical form $\sqcup_{i \in I} \sqcap_{j \in J} a_{ij}$ is equal to $\sqcap_{f \in J^I} \sqcup_{i \in I} a_{i, f(i)}$. In what follows, we will mostly refer to the canonical form of elements of $\mathcal{M}^+$, not to its dual form. Meinongian predication can be characterized more closely by:

M⁺⁺52 $\quad \sqcup_{i \in I} a_i \in b \equiv \exists_{i \in I} a_i \in b$

which is a general form of thesis **M18**.

4. Consistency and Completeness of M-objects

$\mathcal{M}^+$ is in a sense a *weakly extensional* lattice. Every M-object of the lattice is determined by a set of universal M-objects (properties). In $\mathcal{M}^+$ we can define noncontradictory M-objects as fulfilling the condition $\bar{a} \not\in b$ (cf. **df-noncontrad₀**). **M19** and **M⁺34** can be put together:

M⁺⁺53 $\quad a \leq \bar{a} \equiv \forall b, c (a \in b \land a \in c \supset \bar{b} \in c)$

and, accordingly, we may understand consistent M-objects as possessing only noncontradictory properties:

df-con₁ $\quad \text{con}_1(a) \equiv_{df} \forall b, c(a \in b \land a \in c \supset \bar{b} \in c)$

Noncontradictory properties can also be defined by appeal to the notion of atom. Atoms are consistent and complete M-objects (cf. **M20**). So, for any pair of properties, if they are possessed by an individual M-object then they will be noncontradictory:

df-noncontrad₁ $\quad \text{noncontrad}_1(a,b) \equiv_{df} \exists d\, (\mathcal{A}(d) \land d \in b \land d \in c))$

An object will be considered as consistent if it does not possess a pair of contradictory properties, therefore the consistency of M-objects may be defined in the following way:

---

[18] Sets of indices may be empty.

df-con$_2$ $\quad$ con$_2(a) \equiv_{df} \forall b, c(a \in b \land a \in c \supset \exists d\, (\mathcal{A}(d) \land d \in b \land d \in c))$

The noncontradiction may also be introduced in a more familiar and straightforward way. Let a be a universal M-object, i.e. $u(a)$. Then, we can define the complement of a in the following way:

(df*) $\quad a^* =_{df} \sqcap\{c: \mathcal{A}(c) \land \neg c \in a\}$[19]

That is, if $\mathcal{A}(c)$ then $c \in a^*$ iff $\neg c \in a$. For instance, if a means *square*, then a^* will mean *non-square*. Let us recall that consistent objects are defined as not having contradictory or complementary properties. Another definition of consistency involves the complementation operation:

df-con$_3$ $\quad$ con$_3(a) \equiv_{df} \forall b(\mathcal{U}(b) \supset (a \in b \supset \neg a \in b^*))$

Since the following biconditionals are true in $\mathcal{M}^+$

M^{++}54 $\quad a \le \bar{a} \equiv \forall b, c(a \in b \land a \in c \supset \exists d\, (\mathcal{A}(d) \land d \in b \land d \in c))$

M^{++}55 $\quad a \le \bar{a} \equiv \forall b(\mathcal{U}(b) \supset (a \in b \supset \neg a \in b^*))$

all four notions of consistency of M-objects: **con$_0$**, **con$_1$**, **con$_2$**, and **con$_3$** turn out to be equivalent in $\mathcal{M}^{++}$.

Similarly to definitions **df-con$_2$** and **df-con$_3$** we may define the notions of completeness of M-objects (as mentioned earlier there is no counterpart of **df-con$_1$**).

df-com$_2$ $\quad$ com$_2(a) \equiv_{df} \forall b, c(\neg a \in b \land \neg a \in c \supset \exists d\, (\mathcal{A}(d) \land \neg d \in b \land \neg d \in c))$[20]

df-com$_3$ $\quad$ com$_3(a) \equiv_{df} \forall b(\mathcal{U}(b) \supset (\neg a \in b \supset \neg a \in b^*))$

Since the following biconditionals are true in $\mathcal{M}^{++}$

M^{++}56 $\quad \bar{a} \le a \equiv \forall b, c(\neg a \in b \land \neg a \in c \supset \exists d\, (\mathcal{A}(d) \land \neg d \in b \land \neg d \in c))$

M^{++}57 $\quad \bar{a} \le a \equiv \forall b(\mathcal{U}(b) \supset (\neg a \in b \supset a \in b^*))$

once again all three notions of completeness of M-objects **com$_0$**, **com$_2$**, and **com$_3$** turn out to be equivalent.

There is one important feature of these equivalences which is worth noting. The equivalence of **con$_0$** and **con$_1$** hold on the ground of $\mathcal{M}^+$ but the rest hold

---

[19] We may also introduce a relation of complementation C which is not restricted to universal M-objects: $\mathsf{C}(a, b) \equiv_{df} \forall c(\mathcal{A}(c) \supset (c \in a \equiv \neg c \in b))$.

[20] Here, the formula $\forall d(\mathcal{A}(d) \supset (\neg d \in b \supset d \in c))$ means that properties represented by b and c are weakly complemented (see ft. 2, cf. also ft.17).

only in $\mathcal{M}^{++}$. This means that con_0 and com_0 are the only notions of consistency and completeness which can be expressed in the basic Meinongian lattice $\mathcal{M}$ and cannot be further explained within the logic. Hence, this lattice lacks more intuitive and familiar notions: con_2, con_3 and com_2, com_3.

Consistency and completeness can be related in a closer way to the notion of atom. A consistent M-object will be such an object which possesses – in the sense of Meinongian predication – only properties of some individual M-object, i.e. some atom. Since in our theoretical framework individual M-objects are consistent according to definitions of con_1, con_2, and con_3, we may suggest that the individuality determines consistency and, what follows, we may propose another definition of consistency:

df-con$_4$ $con_4(a) \equiv_{df} \exists c (\mathcal{A}(c) \wedge c \Subset a)$

con_4 is stronger than con_1, con_2, and con_3 (cf. M23), and essentially stronger at that point (see a counterexample below).

In an analogous way, we may introduce a new notion of completeness. Since individuals are complete according to definitions of com_1, com_2, and com_3, we may propose the following definition of completeness:

df-com$_4$ $com_4(a) \equiv_{df} \exists c (\mathcal{A}(c) \wedge a \Subset c)$

As before, com_4 is stronger than com_1, com_2, and com_3 (cf. M24) and essentially stronger at that point. To see this, let us assume that our lattice contains only three atoms: c_1, c_2, c_3 and consider the following M-object:

$a = (c_1 \sqcap c_2) \sqcup (c_1 \sqcap c_3) \sqcup (c_2 \sqcap c_3) \sqcup (c_1 \sqcap c_2 \sqcap c_2)$.

a is consistent and complete in the sense of con_0 (and con_1, con_2, con_3) and com_0 (and com_2, com_3) respectively. However, a is neither con_4 nor com_4, which means that it cannot be identified with any individual M-object (atom).[21]

We have not yet mentioned anything about the notion of existence of M-objects. Certainly, this existence may be associated with the individuality of M-objects[22], but the ontology of Meinong's provenience allows a wider notion of existence than that of individuality. The M-object *black dog* conceived as an object constituted of exactly two properties *being black* and *being a dog* is not an individual but rather

---

21 Dale Jacquette even claims that maximal consistency. i.e. consistency and completeness is a sufficient condition for the existence of individual, cf. Jacquette (2002, ch. 2). Our example shows that this is not the case.
22 Although, generally, not every individual must exist – there can be merely possible individuals. However, our approach does not accommodate modal framework.

an incomplete M-object (e.g. it is neither dachshund nor not-dachshund). Hence, it can be said that the object exists because it is true that there exist black cats. What makes such M-object existent, is the fact that there is at least one individual M-objects possessing all properties possessed by that M-object (again, we assume here that all individuals are existent). So, we may define existence in the same way as we have defined **con**$_4$:

df-exist $\text{exist}(a) \equiv_{df} \exists c(\mathcal{A}(c) \wedge c \Subset a)$

5. Set-theoretical Meinongian Lattice

Now we are going to discuss a set-theoretical structure that is proven to be an atomic Meinongian lattice. Let D be any set (possibly empty) and let us consider the following structure $\langle \mathbb{M}(D), \cap, \cup, \sim \rangle$ where:

$\mathbb{M}(D) = \{a \in \wp(\wp(D)): \text{if } X \in a \text{ and } X \subseteq Y \subseteq D \text{ then } Y \in a\}$,

and $\cap, \cup$ are ordinary operations on sets: intersection and union respectively.[23] It is worth emphasizing that members of $\mathbb{M}(D)$ are closed under supersets. For convenience, we also define a closure operation $[\![\ ]\!]$ such that for every $a \in \wp(\wp(D))$,

$[\![a]\!] = \{X \subseteq D: (\exists Y \in a)(Y \subseteq X)\}$.[24]

Operation $\sim$ is defined as follows:

df$\sim$ $\tilde{a} =_{df} \{X \subseteq D: (\forall Y \in a)(X \cap Y \neq \emptyset)\}$

Theorem 5.1 The structure $\langle \mathbb{M}(D), \cap, \cup, \sim \rangle$ is De Morgan lattice.

Proof. Operations $\cap, \cup$ obviously satisfy distributive laws. Therefore, it suffices to prove that $\sim$ is De Morgan lattice operation, i.e. it fulfills **DM1** and **DM2**.

$X \in \tilde{\tilde{a}} \equiv (\forall Y \in \tilde{a})(X \cap Y \neq \emptyset)$

$(\forall Y)(Y \in \tilde{a} \supset X \cap Y \neq \emptyset)$

$\equiv (\forall Y)((\forall Z \in a)(Z \cap Y \neq \emptyset) \supset X \cap Y \neq \emptyset)$

$\equiv X \in a$

The last biconditional needs some explanation. Let us assume that $X \in a$ and for any Y, let $(\forall Z \in a)(Z \cap Y \neq \emptyset)$. Then $X \cap Y \neq \emptyset$.

---

[23] Notice that set-theoretical operations on $\mathbb{M}(D)$ yield elements of $\mathbb{M}(D)$.
[24] Clearly, $\mathbb{M}(D) = \{[\![a]\!]: a \in \wp(\wp(D))\}$.

Now suppose that $(\forall Y)((\forall Z \in a)(Z \cap Y \neq \emptyset) \supset X \cap Y \neq \emptyset)$ holds but $X \notin a$. Then for every $Z \in a$, $Z - X \neq \emptyset$. It also follows that $Z \cap (Z - X) \neq \emptyset$ and, by the assumption,

$X \cap (Z - X) \neq \emptyset$. Contradiction (for $X \cap (Z - X) \neq \emptyset$).

To prove the contraposition, notice that according to **df**$^\sim$ if $a \subseteq b$ then:

$X \in \tilde{b} \equiv (\forall Y \in b)(X \cap Y \neq \emptyset)$

$\supset (\forall Y \in a)(X \cap Y \neq \emptyset)$

$\equiv X \in \tilde{a}$

The proof of the reverse implication goes as follows. First, according to the double negation for $^\sim$ the following equivalence is yielded from **df**$^\sim$: $X \in a \equiv (\forall Y \in \tilde{a})(X \cap Y \neq \emptyset)$. Thus if $\tilde{b} \subseteq \tilde{a}$ then:

$X \in a \equiv (\forall Y \in \tilde{a})(X \cap Y \neq \emptyset)$

$\supset (\forall Y \in \tilde{b})(X \cap Y \neq \emptyset)$

$\equiv X \in b$. ∎

Let us define an additional operation in $\langle \mathbb{M}(D), \cap, \cup, ^\sim \rangle$:

dfφ $\varphi(a) =_{df} \{X \subseteq D : \cap a \subseteq X\}$

We can easily notice that if we interpret φ as the filtration function then, accordingly, the Meinongian predication can be defined as follows:

df$\sqsubset$ $a \sqsubset b \equiv_{df} \varphi(b) \subseteq a$[25]

Theorem 5.2 The structure $\mathcal{M}(D) = \langle \mathbb{M}(D), \cap, \cup, ^\sim, \varphi \rangle$ is an atomic Meinongian lattice where φ is the filtration operation and $\{[\![d]\!] : d \in D\}$ is the set of atoms.

Proof. First, we must show that $\mathcal{M}5$-$\mathcal{M}7$ hold in $\mathcal{M}(D)$. Obviously, set inclusion is the ordering of the lattice. Therefore, $\mathcal{M}5$ is satisfied for $a \subseteq \varphi(a)$. $\mathcal{M}6$ follows from equality $\varphi(a \cap b) = \varphi(a) \cap \varphi(b)$ and $\mathcal{M}7$ from $\varphi(\varphi(a)) = \varphi(a)$.

Now, let us consider $\mathcal{M}^+29$. Suppose that $\forall c(a \sqsubset c \supset b \sqsubset c)$, i.e. by **df**$\sqsubset$, $\forall c(\varphi(c) \subseteq a \supset \varphi(c) \subseteq b)$ and $\forall c(\cap c \in a \supset \cap c \in b)$. But since for every $\cap c \in \wp(D)$ there exists $d \in \wp(D)$, then the last condition means $\forall d(d \in a \supset d \in b)$ and by extensionality of sets: $a \leq b$.

---

25 I.e. $a \sqsubset b \equiv_{df} (\exists A \in a)(A \subseteq \cap b)$.

To prove $\mathcal{M}^{++}35$, let us notice first that for $d \in D$, $[\![\{\{d\}\}]\!] = \{X \subseteq D: d \in X\}$ is an atom. Now suppose that $\varphi(a) \subseteq \widetilde{\varphi(b)}$, i.e. $\cap a \in \widetilde{\varphi(b)}$. By $\mathbf{df}^\sim$, $\cap a \cap \cap b \neq \emptyset$. So there exists $d \in D$ such that $d \in \cap a \cap \cap b$. Hence $d \in \cap a$ and $d \in \cap b$. What follows, $\{X \subseteq D: d \in X\} \sqsubset \varphi(a)$ and $\{X \subseteq D: d \in X\} \sqsubset \varphi(b)$. ∎

Theorem 5.3 Every $\mathcal{M}^{++}$ lattice is isomorphic to $\mathcal{M}(D)$ for some D.

Proof. Let $\mathcal{M}^{++} = \langle \mathbf{M}, \sqcap, \sqcup, \bar{}, f \rangle$ be an atomic lattice and $\mathcal{A}$ be the set of its atoms. We will show that $\mathcal{M}^{++}$ is isomorphic to $\mathcal{M}(\mathcal{A}) = \langle \mathbb{M}(\mathcal{A}), \cap, \cup, \sim, \varphi \rangle$. Let a be any element of $\mathbf{M}$ displayed in its canonical form: $a = \sqcup_{i \in I} \sqcap_{j \in J} a_{ij}$

We define mapping h from $\mathbf{M}$ to $\mathbb{M}(\mathcal{A})$ in the following way:

$h(a) = [\![\{\{a_{ij} \in \mathcal{A}: j \in J\}: i \in I\}]\!]$,

where $h(\Omega) = [\![\emptyset]\!] = \wp(D)$, $h(\mho) = \emptyset$ ($[\![\emptyset]\!]$ is the greatest element and $\emptyset$ is the least element of $\mathbb{M}(\mathcal{A})$). It is clear that h is one-to-one. We will show that:

a) $h(a \sqcup b) = h(a) \cup h(a)$
b) $h(a \sqcap b) = h(a) \cap h(a)$
c) $h(\bar{a}) = \widetilde{h(a)}$
d) $h(f(a)) = \varphi(h(a))$

For simplicity, we will consider only sample finite canonical forms of $\mathcal{M}$-objects which will not diminish the generality of the proof. Let us suppose that $a = (a_1 \sqcap a_2) \sqcup (a_3 \sqcap a_4)$ and $b = b_1 \sqcup (b_2 \sqcap b_3 \sqcap b_4)$. So $h(a) = [\![\{\{a_1, a_2\}, \{a_3, a_4\}\}]\!]$, $h(b) = [\![\{\{b_1\}, \{b_2, b_3, b_4\}\}]\!]$.

Ad a) $h(a \sqcup b) = h((a_1 \sqcap a_2) \sqcup (a_3 \sqcap a_4) \sqcup b_1 \sqcup (b_2 \sqcap b_3 \sqcap b_4))$

$= [\![\{\{a_1, a_2\}, \{a_3, a_4\}, \{b_1\}, \{b_2, b_3, b_4\}\}]\!] = [\![\{\{a_1, a_2\}, \{a_3, a_4\}\}]\!] \cup [\![\{\{b_1\}, \{b_2, b_3, b_4\}\}]\!]$

$= h(a) \sqcup h(a)$

Ad b) $h(a \sqcap b) = h(((a_1 \sqcap a_2) \sqcup (a_3 \sqcap a_4)) \sqcap (b_1 \sqcup (b_2 \sqcap b_3 \sqcap b_4)))$

$= h((a_1 \sqcap a_2 \sqcap b_1) \sqcup (a_3 \sqcap a_4 \sqcap b_1) \sqcup (a_1 \sqcap a_2 \sqcap b_2 \sqcap b_3 \sqcap b_4) \sqcup (a_3 \sqcap a_4 \sqcap b_2 \sqcap b_3 \sqcap b_4))$

$= [\![\{\{a_1, a_2, b_1\}, \{a_3, a_4, b_1\}, \{a_1, a_2, b_2, b_3, b_4\}, \{a_3, a_4, b_2, b_3, b_4\}\}]\!]$

$= [\![\{\{a_1, a_2\}, \{a_3, a_4\}\}]\!] \cap [\![\{\{b_1\}, \{b_2, b_3, b_4\}\}]\!]$

$= h(a) \cap h(a)$

Ad c) $h(\bar{a}) = h(\overline{(a_1 \sqcap a_2) \sqcup (a_3 \sqcap a_4)}) = h((\bar{a}_1 \sqcup \bar{a}_2) \sqcap (\bar{a}_3 \sqcup \bar{a}_4))$

$= h((a_1 \sqcup a_2) \sqcap (a_3 \sqcup a_4)) = h((a_1 \sqcap a_3) \sqcup (a_1 \sqcap a_4) \sqcup (a_2 \sqcap a_3) \sqcup (a_2 \sqcap a_4))$

$= [\![\{\{a_1, a_3\}, \{a_1, a_4\}, \{a_2, a_3\}]\!], \{a_2, a_4\}\}\!] = \{X: X \cap \{a_1, a_2\} \neq \emptyset \; i \; X \cap \{a_3, a_4\} \neq \emptyset\}$

$= \widetilde{h(a)}$

Ad d) Since $f(a) \in f(a)$ then, by **M**⁺21, for some set X of atoms: $f(a) = \sqcap X$. Assuming that $a = \sqcup_{i \in I} \sqcap_{j \in J} a_{ij}$, it is easy to notice that:

$$X = \{ b \in \mathcal{A}: (\forall i)(\exists j) b = a_{ij}\} \text{ and } \cap\{\{a_{ij} \in \mathcal{A}: j \in J\}: i \in I\} = X$$

So $h(f(a)) = h(\sqcap X) = [\![\{X\}]\!] = [\![\{\cap h(a)\}]\!] = \varphi(h(a))$. ∎

Theorem 5.4 There is a homomorphism from $\langle \mathbf{M}, \sqcap, \sqcup, \mho, \Omega \rangle$ to $\langle \mathcal{U}, \sqcap, \uplus, \sqcap, \mathcal{A}, \Omega \rangle$, where $\mathcal{U}$ is the set of universal objects and $a \uplus b = f(a \sqcup b)$[26].

Proof. We will show that the filtration function is the homomorphism in question. Let $a, b \in \mathbf{M}$ Then:

$f(a \sqcap b) = f(a) \sqcap f(b)$

$f(a \sqcup b) = f(f(a) \sqcup f(b)) = f(a) \uplus f(b)$

The last identity follows from **M**17. Clearly, $f(\mho) = \sqcap \mathcal{A}$. ∎

Theorem 5.5 The structure $\mathcal{U} = \langle \mathcal{U}, \sqcap, \uplus, *, \sqcap \mathcal{A}, \Omega \rangle$, where $\mathcal{U}$ is the set of universal M-objects, is Boolean algebra with $\mathcal{A}$ as the set of its atoms.

Proof. It is enough to show that for every $a \in \mathcal{U}$:

(1) $a \sqcap a^* = \sqcap \mathcal{A}$
(2) $a \uplus a^* = \Omega$

For $X \subseteq \mathcal{A}$ let $a = \sqcap X$ and $a^* = \sqcap(\mathcal{A} - X)$. Then for any $c \in \mathcal{A}$:

$c \in a \sqcap a^* \equiv c \in a \vee c \in a^*$

$\equiv c \in \sqcap X \vee c \in \sqcap (\mathcal{A} - X)$

$\equiv (\exists b \in X)(c \in b) \vee (\exists b \in \mathcal{A} - X)(c \in b)$

---

[26] The operation $\sqcup$ is not an inner operation on $\mathcal{U}$ and that is why we restrict it to the operation $\uplus$. $\mho, \Omega$ and $\sqcap \mathcal{A}, \Omega$ are the greatest and the least elements of the two lattices.

$\equiv (\exists b \in X \vee b \in \mathcal{A} - X)(c \Subset b)$

$\equiv (\exists b \in \mathcal{A})(c = b)$

$\equiv c \in \mathcal{A}$

$\equiv c \Subset \sqcap \mathcal{A}$

Since $\mathcal{U}(X)$ and $\mathcal{U}(\mathcal{A} - X)$, the identity (1) follows by **M⁺9**.

Now for any $c \in \mathcal{A}$:

$c \Subset \Omega \equiv c \Subset \sqcap \emptyset$

$\equiv (\exists b \in \emptyset)(c \Subset b)$

$\equiv (\forall b \in X \cap \mathcal{A} - X)(c \Subset b)$

$\equiv (\forall b \in X \wedge b \in \mathcal{A} - X)(c \Subset b)$

$\supset (\forall b \in X)(c \Subset b) \wedge (\forall b \in \mathcal{A} - X)(c \Subset b)$

$\equiv c \Subset \sqcap X \wedge c \Subset \sqcap (\mathcal{A} - X)$

$\equiv c \in a \wedge c \in a^*$

$\equiv c \in a \sqcup a^*$

$\equiv c \Subset f(a \sqcup a^*)$

$\equiv c \Subset a \Cup a^*$

From **M⁺8** it follows that $\Omega \leq a \Cup a^*$. On the other hand, since Ω is the greatest element, $a \Cup a^* \leq \Omega$. Hence $a \Cup a^* = \Omega$. ∎

6. Closing Remarks. Philosophical Interpretations

We have proposed an algebraic interpretation of Meinongian theory of objects in several steps. First, we have introduced a basic Meinongian lattice $\mathcal{M}$, then $\mathcal{M}^+$, and finally the strongest $\mathcal{M}^{++}$. As we have noticed, $\mathcal{M}$ is essentially quantifier free, i.e. any thesis involving quantifiers is logically equivalent to one without quantifiers. It is possible to express in $\mathcal{M}$ notions of consistency and completeness of M-objects by means of definitions **df-con**$_0$ and **df-com**$_0$. However, these notions have no further explanation, i.e. they are primitive concepts in the sense that there is no justification for why some M-objects are consistent or complete. Strengthening $\mathcal{M}$ to a complete lattice accomplishes one of the basic principles of Meinong's theory of objects, i.e., the so called *Annahmen* thesis, which says that every set (class) of properties constitutes exactly one object. According to **df⊔**, for every set of universal M-objects.

i.e. properties there is an M-object which possesses – in the sense of Meinongian predication ∈ – exactly those properties. When $\mathcal{M}^+29$ is added and consequently $\mathbf{M}^+33$ holds, the lattice $\mathcal{M}^+$ will fulfil another principle of the theory, namely the identity principle which says that objects possessing the same properties are identical. Still, in $\mathcal{M}^+$ only one notion of noncontradiction is available, i.e. **noncontrad**$_0$, which can be used to define the consistency of M-objects. If we add $\mathcal{M}^{++}35$ to $\mathcal{M}^+$, we will obtain an atomic Meinongian lattice $\mathcal{M}^{++}$ with full identification of universal M-objects (properties) by means of atoms (individual M-objects). Roughly speaking, properties are identical if they are possessed by the same individuals. Consequently, noncontradiction and complementarity are defined in a more usual way and, what follows, also consistency and completeness of M-objects receive the meaning closer to the one usually associated with these notions.

Conspicuously, the stronger a Meinongian lattice is, the more extensional it becomes. The basic lattice $\mathcal{M}$ may be considered as the most intesional one. M-objects in $\mathcal{M}$ may be nonidentical even if they possess the same properties (i.e. universal M-objects). $\mathcal{M}^+$ is extensional in the sense that every M-object can be fully determined by a set of properties, which is guaranteed by $\mathbf{M}^+33$. However, properties themselves are not identified by sets of individual M-objects (atoms). Only $\mathcal{M}^{++}$ may be considered as fully extensional: every M-object is correlated with the set of properties and every property is correlated with individual M-objects. Extensionality is clearly exhibited by the set-theoretical lattice $\mathcal{M}(\mathbf{D})$ which is isomorphic to $\mathcal{M}^{++}$ (see **Theorem 5.3**). One can easily verify that by $\mathbf{df} \sqsubset$ the following predications are true in $\mathcal{M}(\mathbf{D})$:

{set of circles, set of squares} ⊏ {set of circles}
{set of circles, set of squares} ⊏ {set of squares}
{set of circles, set of squares} ⊏ {set of geometrical figures}
{set of circles, set of squares} ⊏ {set of non-circles}
{set of circles, set of squares} ⊏ { set of non-squares}

and they make sentences (I)–(V) from Section 1 true. At the same time, the following predications are false by the definition:

{set of circles, set of squares} ⊏ {set of black things}
{set of circles, set of squares} ⊏ {set of non-black things}

which correspond to sentences (VII) and (VIII) respectively. It is worth noting that the sentence (VI) represented by

{set of circles, set of squares} ⊏ {set of circles, set of squares}, or by
{set of circles, set of squares} ⊏ {set of round squares}

is not true. Although this may seem to be counterintuitive, it is the cost of assuming that possessing by an object a conjunction of properties (i.e. *being round square*) is stronger than possessing these properties separately (*being round* and *being a square*). Of course,

{**set of round squares**} ⊏ {**set of round squares**}

is true – but it is trivially true, because M-object {**set of round squares**} is identical with {∅} and as such it possesses all properties.

We now turn back to the example of M-object a given in Section 4 and interpret the object in $\mathcal{M}(D)$. So let $D = \{c_1, c_2, c_3\}$ and $a = \{\{c_1, c_2,\}, \{c_1, c_3,\} \{c_2, c_3,\} \{c_1, c_2, c_3\}\}$. The object is consistent and complete (maximally consistent) but cannot be identified with any member of D – it cannot be treated as an individual. a is not strongly consistent and strongly complete, **con**$_4$ and **com**$_4$ respectively. In particular, it does not exist in the sense of **exist**. This result conforms the view that formal-ontological notions like consistency and completeness cannot entail metaphysical notion of existence.

Acknowledgments

I would like to thank Marcin Wolski for his valuable comments on the earlier draft of this paper.

References

Berto, F. (2013). *Existence as a Real Property. The Ontology of Meinongianism*. Dordrecht: Springer Verlag.

Dunn, J. M. and Hardegree, G. M. (2001). *Algebraic Methods in Philosophical Logic*. Oxford: Clarendon Press,.

Jacquette, D. (1996). *Meinongian Logic: The Semantics of Existence and Nonexistence*. Berlin and New York: Walter de Gruyter & Co.

Jacquette, D. (2002). *Ontology*. Montreal: McGill Queen's University Press.

Jacquette, D. (2015). *Alexius Meinong. The Shephard of Non-Being*. Dordrecht: Springer Verlag.

Parsons, T. (1980). *Nonexistent Objects*. New Haven and London: Yale University Press.

Paśniczek, J. (1996). Leśniewski's Ontology vs. Meinongian Ontology. *Axiomates*, 1–2.

Paśniczek, J. (1998). *The Logic of Intentional Objects. A Meinongian Version of Classical Logic*. Dordrecht/London/Boston: Kluwer.

Paśniczek, J. (2014a). *Predykacja. Elementy ontologii formalnej przedmiotów, własności i sytuacji.* Kraków: Copernicus Center Press.

Paśniczek, J. (2014b). Toward a Meinongian Calculus of Names. In: M. Antonelli, M. David (eds.), *Logical, Ontological and Historical Contributions on the Philosophy of Alexius Meinong. Meinong Studies.* Berlin and Boston: De Gruyter.

Perszyk, K. J. (1993). *Nonexistent Objects. Meinong and Contemporary Philosophy.* Nijhoff: Kluwer.

Routley, R. (1980). *Exploring Meinong's Jungle and Beyond.* Department Monograph #3, Philosophy Department, Research School of Social Sciences, Canberra, Australian National University.

Swanson, C. (2011). *Reburial of Nonexistents. Reconsidering Meinong-Russell Debate.* Amsterdam and New York: Rodopi.

Zalta, E. (1983). *Abstract Objects: An Introduction to Axiomatic Metaphysics.* Dordrecht: D. Reidel.

Zalta, E. (1988). *Intensional Logic and the Metaphysics of Intentionality.* Cambridge, MA and London.

Zalta, E. (1999). *Principia Metaphysica.* http://mally.stanford.edu/.

Index

A
accidental predication, 14, 16, 25–34, 37–39, 41, 45–49
acts of predication, 199–201, 204–209, 211–218, 223, 226, 228
Aristotle, 9, 13, 14, 16, 25–49, 102–107, 110, 114, 115, 229

C
class, 17, 77–80, 82, 83, 85, 86, 88, 89, 93–98, 118, 119, 121, 141, 142, 160, 169, 170, 225, 230, 231, 237, 248, 250, 252, 253, 267
cognitive capacities, 53–55, 57, 62–65, 83, 98
combinatorial, 17, 101, 109
complementary properties, 247–249, 253, 257, 261
complete lattice, 247, 258, 267
concept, 16, 18, 25, 36, 53–55, 57, 58, 62–70, 72, 73, 75, 77, 78, 82, 83, 89, 93, 94, 96, 98, 119, 130, 139, 141, 147, 155, 156, 159, 167, 169, 171, 172, 174, 178, 182, 184, 189, 229, 254, 267
concept *horse*, 18, 155, 156, 167–171, 174–184
Conceptualism, 53, 57, 58
conceptualist logic, 17, 53, 54, 57, 68, 69, 75, 83
contradictory properties, 247–250, 253, 257, 260, 261
conventionalist, 18, 145, 146, 149, 152–154, 156–158, 160, 162
copula, 15, 17, 53, 57, 70, 71, 73, 81, 87, 88, 147–149, 154, 156, 158, 159, 176, 184, 185 228, 229
correctness, 199, 201, 206, 207–218, 243

D
De Morgan lattice, 247, 250, 251, 254, 255, 263
De Morgan negation, 247, 250, 253, 254
Dummett, Michael, 17, 18, 119, 129–140, 148, 171
dyadic relation, 101, 109, 111, 112, 114, 118, 122, 123, 149

E
essential predication, 14, 16, 25–37, 39
existence, 55, 58, 113, 114, 247, 249, 257, 262, 263, 269

F
filtration function, 247, 250, 255, 256, 264, 266
formalism, 17, 101, 107, 108
Frege, Gottlob, 12, 13, 15–18, 55, 62, 63, 98, 106, 114–119, 129–131, 133–137, 141, 145–150, 152–159, 162, 163, 167–174, 176–183, 185, 188, 192, 193, 195, 200, 202, 224, 225, 228, 229

G
Geach, Peter, 17, 18, 57–59, 77, 129–131, 133–141, 156, 157
greatest element, 247, 258, 265, 267

H
Husserl, Edmund, 63, 141, 142

I
(in)complete objects (M-objects), 19, 247, 249, 252

(in)consistent objects (M-objects), 247–250, 252, 253, 261
individuals (individual M-objects), 247, 257, 260, 262, 263, 268

K
Kripke, Saul, 207, 210–212, 215

L
least element, 247, 258, 265, 266
Łukasiewicz, Jan, 42, 44–49

M
mass nouns, 16,17, 53, 54, 59, 60, 73–75, 79, 83–89, 188
Medieval Logic, 53, 56, 59, 61, 64, 70
Meinong, Alexius, 19, 68, 247–252, 262, 267
Meinong's theory of objects, 19, 247, 248, 250–252, 267
Meinongian lattice, 247, 250, 255, 257, 258, 262–264, 267, 268
Meinongian predication, 16, 19, 247, 251–256, 260, 262, 264, 268
mental act, 10, 16, 17, 53–56, 62, 63, 66, 68, 70, 73, 76, 79, 85, 87, 89, 204, 209, 212
Mill, John Stuart,
M-objects (Meinongian objects), 247, 248, 250–253, 255–257, 259–263, 265–268
Montague, Richard, 18, 66–68, 167, 168, 174, 179, 181, 224

O
objectivity, 199, 207, 208, 213, 215–217
Ockham, Willaim, 56–58, 70, 105
orders of classes, 93, 95, 98

P
Platonism, 19, 199, 206–208, 210, 212, 215–217

pluralities, 53, 73, 74, 76–79, 82, 83, 85, 86, 89
Posterior Analytics, 16, 25, 26, 38, 40
predication of accidents, 16, 25, 26, 32, 36, 38–41
predication of essence, 16, 25, 37
procedural semantics, 223, 224, 237
properties, 10, 19, 34, 63, 65, 109, 112, 118, 119, 121, 122, 133, 137, 141, 142, 150, 156, 158, 160, 161, 163, 181, 189, 193, 194, 199–201, 204, 218, 223, 225–227, 229, 230, 232, 233, 237, 238, 240, 241, 247–253, 255–257, 260–263, 267–269

Q
Quine, Willard Van Orman, 9, 10, 43, 111, 117, 122

R
reference, 9, 16–18, 53, 54, 56–58, 60, 66, 68, 69, 73, 75, 76, 79, 83–85, 87, 89, 116, 129, 131–135, 138, 155–157, 159, 167, 169, 171, 173, 174, 176, 179–181, 189, 191, 193, 194,
rule-following, 18, 19, 54, 63, 199, 201, 206–208, 212, 213, 218
Russell, Bertrand, 61, 64, 74, 77, 78, 80, 111, 114, 116, 122, 155–157, 191, 202, 211, 212, 225–227, 229–231, 233

S
semantics, 12, 13, 15–18, 56, 77–79, 85, 86, 89, 114, 116, 129, 137–141, 167–170, 173, 174, 176, 185, 191, 194, 223–225, 229, 236, 237
situation semantics, 18, 129, 139–141
speech act, 10, 16, 17, 53, 55, 61, 62, 64, 66, 68, 70, 73, 76, 79, 80, 82, 85, 87, 89, 103, 106, 117, 160, 200, 228, 235, 236

Stalnaker, Robert, 199, 201, 205–207
Sulzer, Johann Caspar, 17, 93, 95, 97–99

T
type-theoretic semantics, 18, 167, 168, 174, 176
type-theory, 18, 68, 167, 168, 170, 174–177, 179, 180, 183, 184, 190–194, 227, 237, 239, 242, 243

U
unity of the proposition, 11, 13, 18, 117, 129, 137, 145, 146, 148, 152, 153, 160, 162, 171, 199, 202, 225–227, 229, 232, 235
unity of the sentence, 18, 145, 146, 148, 151–153, 156, 158, 159, 160, 162
universals (universal M-objects), 256, 257, 260, 261, 266, 267
unsaturatedness, 18, 53, 136, 139, 145–148, 150, 152, 154, 156, 157, 159, 162

W
Wittgenstein, Ludwig, 19, 141, 199, 201, 207, 210–212, 214, 215, 218

Studies in Philosophy of Language and Linguistics
Edited by Piotr Stalmaszczyk

Vol. 1 Piotr Stalmaszczyk / Luis Fernández Moreno (eds.): Philosophical Approaches to Proper Names. 2016.

Vol. 2 Piotr Stalmaszczyk (ed.): Philosophical and Linguistic Analyses of Reference. 2016.

Vol. 3 Martin Hinton (ed.): Evidence, Experiment and Argument in Linguistics and the Philosophy of Language. 2016.

Vol. 4 Piotr Stalmaszczyk (ed.): From Philosophy of Fiction to Cognitive Poetics. 2016.

Vol. 5 Luis Fernández Moreno: The Reference of Natural Kind Terms. 2016.

Vol. 6 Szymon J. Napierała: Symmetry Breaking and Symmetry Restoration. Evidence from English Syntax of Coordination. 2017.

Vol. 7 Piotr Stalmaszczyk (ed.): Philosophy and Logic of Predication. 2017.

www.peterlang.com

www.ingramcontent.com/pod-product-compliance
Ingram Content Group UK Ltd.
Pitfield, Milton Keynes, MK11 3LW, UK
UKHW041902230426
12049UKWH00002B/16